THE
LAST
SORCERERS

ALSO BY RICHARD MORRIS

Achilles in the Quantum Universe: The Definitive History of Infinity

Artificial Worlds: Computers, Complexity, and the Riddle of Life

Big Questions: Probing the Promise and Limits of Science

Cosmic Questions: Galactic Halos, Cold Dark Matter and the End of Time

Dismantling the Universe: The Nature of Scientific Discovery

The Evolutionists: The Struggle for Darwin's Soul

The Nature of Reality: The Universe After Einstein

The Universe, the Eleventh Dimension, and Everything: What We Know and How We Know It

Time's Arrows: Scientific Attitudes Toward Time

THE
LAST
SORCERERS

THE PATH FROM ALCHEMY
TO THE PERIODIC TABLE

Richard Morris

Joseph Henry Press

Washington, D.C.

Joseph Henry Press • 500 Fifth Street, NW • Washington, DC 20001

The Joseph Henry Press, an imprint of the National Academies Press, was created with the goal of making books on science, technology, and health more widely available to professionals and the public. Joseph Henry was one of the founders of the National Academy of Sciences and a leader in early American science.

Any opinions, findings, conclusions, or recommendations expressed in this volume are those of the author and do not necessarily reflect the views of the National Academy of Sciences or its affiliated institutions.

Library of Congress Cataloging-in-Publication Data

Morris, Richard, 1939-2003
  The last sorcerers : the path from alchemy to the periodic table / Richard Morris.
    p. cm.
Includes bibliographical references and index.
  ISBN 0-309-08905-0 (hbk.)—ISBN 0-309-50593-3 (PDF)
  1. Chemistry—History. I. Title.
  QD11.M86 2003
  540'.9—dc22

                                 2003014790

Printed in the United States of America

First Printing, October 2003
Second Printing, March 2004

CONTENTS

Preface		vii
1	The Four Elements	1
2	Prelude to the Birth of Chemistry	26
3	The Sceptical Chymist	45
4	The Discovery of the Elements	68
5	A Nail for the Coffin	90
6	"Only an Instant to Cut Off That Head"	108
7	The Atom	130
8	Problems with Atoms	145
9	The Periodic Law	157
10	Deciphering the Atom	176
Epilogue: The Continuing Search		203
Appendix: A Catalog of the Elements		223
Further Reading		261
Index		265

PREFACE

Nowadays we hear a great deal about physicists' ongoing effort to understand the nature of the universe's ultimate constituents. Numerous books are written about the physics of elementary particles, about the hypothetical objects known as superstrings, and about the "dark matter" that constitutes a large part of the universe's mass. Millions of words are written about attempts to probe their mysteries.

However, the quest to understand what the world was made of did not begin with discoveries in physics but in the West with the ancient Greeks, who pondered the ultimate constituents of matter and advanced a number of theories before concluding that there were four elements: earth, air, fire, and water. Of course the theory was wrong, but for sheer longevity, it was one of the most successful ever proposed. It lasted more than 2,000 years.

Not until the sixteenth century did questions about the ultimate nature of things began to be asked again. Although the four-element theory continued to be accepted, new attempts were made to better understand it. The questioning continued until the eighteenth century, when the natural philosophers (there were no "scientists" then; that word did not come into general use until the middle of the nineteenth century) who pondered such things created the science of chemistry.

Creating a new science was an arduous task, one that continued over the course of many generations. The four-element theory had held sway for so long that it required well over a century of experimentation, observation, and theorizing to overturn it. One impediment was the lack of a modern conception of a chemical element. And although many chemists believed that matter was made of atoms, they couldn't describe the properties of an atom with any confidence. Some chemists refused to believe that such things even existed. In their view atoms were nothing more than a useful fiction. It wasn't until 1905 that Albert Einstein settled the question, showing that observations of a phenomenon known as the Brownian movement provided proof that atoms were real.

During most of the eighteenth century chemists remained ignorant of the nature of the substances they worked with. None of the three most abundant elements in the Earth's crust—oxygen, silicon, and aluminum—had yet been discovered. Chemists didn't know that air could be broken down into different components, and they remained ignorant of such gasses as oxygen, nitrogen, and hydrogen, all of which play important roles in chemical reactions. Yet they never ceased searching for the key to the universe—knowledge of what the world was made of—and important new discoveries were made in every generation. By the end of the century modern chemistry had been created.

In the course of this surge of chemical research, new problems arose almost as soon as the old ones were solved. By the mid-nineteenth century about 60 chemical elements were known, and it was puzzling that there were so many. Could the universe really have 60 different fundamental components? And if it did, what were the relationships between them? Why did some have properties very much like those of some others? Was it possible to find some order in the chaotic table of the elements?

THE LIVES OF THE CHEMISTS

A number of chemistry histories describe discoveries in great detail. I have not attempted to duplicate what they do so well. Instead I have concentrated on the lives of the people who transformed chemistry into a modern science. I have not shied away from explaining their most important discoveries, but I have not dwelled on the technical details.

The lives of these men were often eventful, but eventful in different ways. For example, there was Robert Boyle, who is widely considered the founder of the science of chemistry. But Boyle was an alchemist as well as a chemist, and he spent the greater part of his life seeking the Philosopher's Stone, the elusive substance that could supposedly transform base metals into gold. His search for the Stone led to some misadventures that I describe in detail.

Today Joseph Priestley is known as one of the discoverers of oxygen.* However, in his own day his liberal political views branded him as a dangerous political radical. Priestley once had to flee when a royalist mob destroyed his house and laboratory, and he later emigrated to the United States when his friends advised him that he was risking his life by remaining in England.

Priestley's contemporary, Henry Cavendish, led an entirely different kind of life. One of the wealthiest men in England, he was a recluse and made his great discoveries in a laboratory he built in his home. Cavendish avoided conversing with men as much as he could, and he fled if he encountered a woman. Once, after running into a maid on a stairway in his house, he had back stairs built for the maids so that he would never have to run into them again.

---

*The Swedish apothecary Carl Wilhelm Scheele also has a strong claim to this distinction. See page 81.

Antoine Lavoisier was a French chemist who had the misfortune to live in revolutionary times, although he was no diehard loyalist. On the contrary, while not politically active, he held views that were very liberal for his day. Lavoisier died on the guillotine. In pre-revolutionary days, he had been a frequent target of diatribes written by the radical leader Jean-Paul Marat. Marat, who once had scientific ambitions, believed that Lavoisier blocked his attempts to gain election to the French Academy of Sciences. Marat was assassinated before Lavoisier was executed, so he played no role in the latter's arrest or trial, but it is significant that he had constantly attacked Lavoisier for his role as a tax farmer. It was for his activities as a tax farmer that Lavoisier was executed.

Dimitri Mendeleev was a bigamist who married a second time after paying an orthodox priest, who was later defrocked, to give him a dispensation. Mendeleev had his long white hair and beard cut only once a year, giving him a somewhat outlandish appearance. However, his contemporaries admired him as the greatest Russian chemist. Though the political situation in Russia caused him problems, he was luckier—if dying at a certain time can be said to be lucky—than Lavoisier in that he passed away quietly before the communist revolution.

Niels Bohr was a physicist, not a chemist. I devote a chapter to his life because he was the scientist who explained why Mendeleev's periodic table had the properties it did. Widely known as a soccer player in his youth, Bohr became the most influential physicist of the first half of the twentieth century. His life, too, was touched by political events. A Jew living in occupied Denmark, Bohr had to flee the country to avoid arrest by the Nazis. In 1939 Bohr discovered a theory that explained nuclear fission, and suggested that uranium 235 could be used to make a bomb. Though he played only a minor role in the American atomic bomb project, Bohr was the first to ponder the political implications of the bomb.

Other major figures discussed in the book include Paracelsus, whose outrageous character and life could not possibly be summarized in a paragraph. Other figures played smaller roles: for example, the German philosopher Gottfried Leibniz, whose interest in alchemy eventually led to his involvement in the production of a new element, phosphorus, from human urine, and Werner Heisenberg, Bohr's friend and director of the German atomic bomb project. The story of the development of chemistry is something like a play in that bit players appear from time to time, contributing to the development of the plot.

The chapters that follow are arranged in chronological order, with the exception of Chapter 4, which discusses the discovery of new chemical elements over the course of two centuries. I thought that I could give a more coherent account if I put that material in a single chapter rather than scattering it throughout the book.

The last chapter, which I have called an epilogue, is also somewhat different from the others. It is a condensed history of twentieth-century particle physics. The search for an understanding of the constituents of matter did not end with Bohr's explanation of the properties of the periodic table after all. On the contrary, the quest continued by being passed from the hands of the chemists into those of the physicists. Because I chose to discuss this material within the framework of a single chapter, I was forced to omit some of the details. However, I think it sufficiently summarizes the paths that the physicists followed once they took on the task of trying to determine what the universe was made of.

I would like to conclude by acknowledging that the idea for this book was not my own. It was suggested to me by Erika Goldman, my former editor at W.H. Freeman and Company. Left to my own devices I wouldn't have thought of telling the tale that I did.

CHAPTER 1

THE FOUR ELEMENTS

For more than two and a half millennia, philosophers and scientists have tried to understand what the universe is made of and the principles on which it operates. Philosophers engaged in this kind of speculation in classical times, and physicists and cosmologists engage in it today. Although their methods are very different, modern scientists share the goal of the ancient philosophers: to find the key to understanding the universe.

The first philosopher to theorize about such matters was Thales of Miletus, at the time, the sixth century B.C., the greatest Greek city Asia Minor. According to Thales there was one fundamental element: water, the material of which everything was made. To the modern mind, such an idea seems absurd. However, it is much more reasonable than it might appear. Lacking evidence to the contrary, it must have seemed very plausible that everything was made of some primal material. And if it was, water was really not a bad candidate.

Thales must have noted that evaporation turned water into mist and that it solidified when it froze. Aristotle suggests that Thales got the idea "perhaps from seeing that the nutriment of all things is moist and kept alive by it. . . . He got his notion from this fact that the seeds of all things have a moist nature, and water is the origin of the nature of moist things." Of course, Aristotle was guessing when he said that. However, it is clear that be believed that Thales had a plausible, if incorrect, idea.

Thales's successor, Anaximander—the exact dates of his birth and death are unknown, but he was said to have been 64 years old in 546 B.C.—agreed that there was one primal material. But he didn't think it was ever encountered on Earth in its pure state. According to Anaximander everything in the world was made of *apeiron*, a substance that was infinite and eternal, and which could take on numerous forms, including those of all the familiar terrestrial materials. "It is neither water nor any of the so-called elements," Anaximander said, "but a nature different from them and infinite, from which arise all the heavens and the worlds within them."

The last of the Miletus philosophers was Anaximenes. His dates are also uncertain, but he must have created his theory before 494 B.C. when the Persians destroyed Miletus. Apparently Anaximenes did not find Anaximander's ideas very convincing, because he maintained that the fundamental element was air. Anaximenes maintained that fire was rarefied air and that air could be condensed into all known substances. Progressive condensations successively condensed it into wind, clouds, water, and finally into earth and stone. "Just as our soul being air, holds us together," he said, "so do breath and air encompass the whole world."

EMPEDOCLES

In classical times stories circulated about Empedocles, a philosopher who lived in Agrigentium, in southern Sicily, around the middle of the fifth century B.C. It was said that he performed miracles, that he could control the winds, and that he had brought a woman who had seemed dead for 30 days back to life. Empedocles was the leader of the democratic party in his native city, and he claimed to be a god. According to legend, he died when he jumped into the crater of Mount Etna in an attempt to prove that he was a god. Whether he actually did this is uncertain, however. According to Diogenes, who related this tale, "Timaeus contradicts all these stories, saying expressly that he departed into Peloponnesus, and never returned at all, on which account the manner of his death is uncertain."

Empedocles made no attempt to create a new theory of matter. Instead, he tried to reconcile the thoughts of his various predecessors. He took Thales's theory that everything was made of water and Anaximenes's idea that the primal substance was air, and added two more elements, earth and fire. Empedocles didn't believe that one kind of matter could be transformed into another. Earth couldn't be changed into water, or water into earth, for example. Thus there had to be more than one element.

Empedocles didn't speak of earth, air, fire, and water as "elements," but as "the roots of all." Nevertheless, each was eternal, and they could be mixed together in varying proportions to produce the substances encountered in the terrestrial world. According to Empedocles, the elements were combined by love and separated by strife. The theory was not as mystical as it sounds. Empedocles seems to have thought of love and strife as physical forces that could act on the particles of matter. "Love," in other words, was a force of attraction.

Empedocles's theory of the four elements was to dominate Western thought for nearly two and a half millennia. It wasn't until the eighteenth century that it was overthrown, because it was endorsed by Aristotle, whose authority was so great that his dogmas often impeded scientific progress. Aristotle added a fifth element, of which the heavenly bodies were supposedly composed. But he agreed with Empedocles that all earthly objects were made of earth, air, fire, and water.

Aristotle elaborated on the theory by assigning qualities to the four elements. Fire was hot and dry, air was hot and moist, water was cold and moist, and earth was cold and dry. This implied that it was possible for one element to be transformed into another. In Aristotle's day it seemed a reasonable theory, one that was supported by common observations. For example, if the "cold" in water were made hot, then the water would be transformed into air. And indeed this is what appears to be happening when water is boiled. When wood was burned, smoke (air), pitch (water), ash (earth), and fire were produced. If two pieces of flint were struck together, a spark was produced that could be used to kindle a fire. Thus it seemed that the fire element must be present in rock.

ALCHEMY

Alchemy was born of a fusion of Greek philosophy and the Egyptian chemical arts in Alexandria, the city founded by Alexander the Great at the mouth of the Nile in 331 B.C. The Egyptians had for centuries practiced embalming, dyeing, glassmaking, and metallurgy, each requiring knowledge of the relevant chemical processes. There were numerous recipes, including ones for making artificial gems and false gold.

Perhaps it was only natural that people steeped in Greek philosophy would think of trying to make gold when they encountered the rich Egyptian tradition of practical chemistry. Hadn't Aristotle said that transformations were possible? Isn't that what happened when, for example, cinnabar (mercury ore) was heated? Heating the red material, cinnabar, caused a pool of liquid metal to form. Didn't other chemical transformations take place when substances were heated, dissolved, melted, filtered, and crystallized?

In one sense, the creation of alchemy represented a step backward. The Egyptians had known seven metallic elements: gold, silver, copper, tin, iron, lead, and mercury, which they associated with the seven planets (Sun, Moon, Mercury, Venus, Mars, Jupiter, and Saturn, respectively). The Greeks, however, failed to recognize them as distinct elements. According to the Aristotelian theory, the metals were mixtures of the traditional four elements. This idea seemed to support the theory that one metal could be transformed into another. All that was needed was to find the chemical procedures that would remove some of one element and add some more of another, or that would change one element into another.

Alexandria was a place where many different religions and cultures encountered one another and where different philosophies flourished. The city was home to Greeks, Egyptians, Jews, and people who had migrated from many places in the Middle East. There were Zoroastrians, Neoplatonists, Mithraists, Christian Gnostics, and adherents of a number of other philosophies and faiths. Alexandria had sober believers in Greek rationalism and also wizards and sorcerers, mystics, astrologers, and prophets.

By A.D. 300, Alexandrian alchemy had become almost entirely mystical, perhaps because the alchemists were influenced by the currents of mystical thought they encountered, perhaps because attempts to transform base metals into gold had failed. One suspects that the latter played at least some role. It is certainly easier to dwell on the idea of the spiritual gold in one's soul than to follow long complex procedures in an attempt to actually make the metal.

At one time scholars believed that the Roman Emperor Diocletian decreed in A.D. 292 that all alchemical books be burned and that the alchemists be expelled from Egypt. But this story is probably apocryphal. At the time, alchemy was unknown in the Roman west. In any case, no decrees were needed. Alexandrian intellectual culture was past its prime by then, and alchemy simply participated in the decline.

After Constantine proclaimed Christianity to be the official cult of the Roman Empire around A.D. 330, the Christians sought to eradicate pagan philosophies, including alchemy. Most likely they would have succeeded if members of a heretical Christian sect, the Nestorians, had not preserved alchemical writings. After Nestorius, the leader of the sect, was excommunicated around A.D. 430, he fled to Syria with his followers. The Nestorians took as many pagan manuscripts and books with them as they could and kept them in the mon-

asteries they founded. Around A.D. 500 the Nestorians were expelled
from Syria. They moved on to Persia, where they founded schools
and translated Hellenistic writings into Syrian. One of the subjects
taught in their schools was alchemy.

ARABIC ALCHEMY

The years 640 to 720 were an era of Muslim conquests. At the end of
the period, the Islamic empire stretched from Spain to Egypt and
from North Africa to Persia. The Muslims engaged in wars of expan-
sion, not in religious war. They didn't seek to convert the peoples they
conquered. Although non-Muslims were taxed, they were permitted
to exercise their religions freely.

Unlike Christians, who wanted to eradicate pagan philosophy,
Muslims had great respect for learning. Muslim rulers patronized
scholars, whatever their religion, and had Greek and Syrian texts
translated into Arabic. Thus Arab scholars gained knowledge of the
thoughts of Plato, Aristotle, and other philosophers and also of
alchemy. It was the Muslims who gave alchemy its name. The word is
derived from the Arabic *alchymia. al* is the Arabic definite article;
words beginning with those two letters, such as alcohol and algebra,
are generally of Arabic origin. The exact meaning and origin of
chymia are uncertain. It used to be thought that it derived from *Khem*,
the ancient name of Egypt. However, recent scholars have cast doubt
on this idea. The Arabs were not much interested in the mystical
accretions that alchemy had acquired. They pursued it in a more
down-to-earth manner, as the early Alexandrian alchemists had done.
Thus, centuries later, alchemy reached Europe as a collection of
chemical recipes and techniques, not a set of esoteric doctrines.

It is in Arabic alchemy that two concepts that were to become
central to European alchemy are encountered for the first time: the
Philosopher's Stone and the elixir of life. The Philosopher's Stone
was a substance reputedly able to transform base metals into gold. In
spite of the name, it wasn't thought of as a stone and was often

described as a "red earth," for example. The elixir of life, as the name implies, was something that could restore youth and prolong life. Presumably it could be made from alchemical gold. But this isn't all there was to alchemy. Its more practical side included procedures to produce dyes and medicines. Since the first stirrings in the alchemist's cauldron, there was always more to alchemy than the quest to make gold.

Arabic alchemy was unknown in the west until the eleventh century when the first translations from Arabic into Latin were made. Two Arab alchemists were especially well known and widely read: Jabir ibn Hayyan, known to Europeans as Geber, and Abu Bakr ibn Zakariyya al-Razi, known as Rhazes. Of more than 2,000 pieces of writing attributed to Jabir, most were compiled by a Muslim religious sect called the Faithful Brethren or Brethren of Purity after he died. The works are written in different styles, which would indicate that they were penned by different authors. The compilation was completed around the year 1000, more than a hundred years after Jabir died. However, it has been established that the work translated into Latin under the title *Summa Perfectionis* was based on translations of Jabir's writing. Thus, although little is known about his life, we know something about the role Jabir played in the evolution of alchemical theory.

Jabir introduced a theory, which was to influence much of later alchemy, that metals were mixtures of sulfur, mercury, and arsenic, except for gold, which was made up of sulfur and mercury alone. The sulfur and mercury of which Jabir spoke were not the substances found in nature. They were purified essences which European alchemists later called "philosophical sulfur" and "philosophical mercury." They were supposed to be quite unlike the common substances. For example, it was said that philosophical sulfur didn't burn. According to Jabir, of all the metals, gold contained the most mercury and the least sulfur. Thus other metals could be transformed into gold if ways were found to increase their mercury content.

We know somewhat more about al-Razi's life than Jabir's. He was a Jewish convert to Islam who became a physician and alchemist in Persia, and he wrote a text on alchemy called *Secret of Secrets*. While the title seems to promise something esoteric, al-Razi's book isn't like that at all; on the contrary, it is a comprehensive and practical laboratory manual that became a valuable tool for European alchemists. The *Secret of Secrets* contains huge lists of chemicals and minerals and comments on their origin. It describes alchemical apparatus, including several kinds of glassware, and chemical techniques. Many of the recipes are stated so clearly that they could easily be followed and carried out in a chemical laboratory today. Unlike most alchemists, al-Razi seems not to have regarded the transmutation of metals as the main goal of alchemy. As a physician, he emphasized the importance to medicine of knowing the chemical substances in medicine. However, the wealth of laboratory techniques described in his book proved invaluable to generations of European alchemists, whatever their goal.

It is said that Al-Razi became blind in his later years, spending them in poverty because he was no longer able to practice medicine. His eyes might have been damaged by chemical fumes. Other stories about him sound a bit fanciful. According to one, a high-ranking individual (said by some to have been the Emir of Khorassan) asked al-Razi to demonstrate a procedure for making gold. When he refused, the Emir lost his temper and struck him on the head with his own book, causing him to become blind. According to another version of the story, the Emir became angry when al-Razi did attempt to make gold but failed. Curiously, this story didn't deter the gold-seeking alchemists who, over a period of centuries, pored over al-Razi's writings.

EUROPEAN ALCHEMY

The appearance of Arabic alchemical works in Latin translation launched European alchemy during the eleventh and twelfth centuries. Although the European alchemists never succeeded in making gold or the elixir of life, they did make some important discoveries. For example, in the early fourteenth century an alchemist known as the False Geber (because he called himself Geber, after his Arab predecessor) discovered how to make strong sulfuric and nitric acids. This was a significant advance. The ancients and the Arabs had known only weak acids, such as acetic acid from vinegar and lactic acid from soured milk. Unlike the weak acids, strong acids are extremely corrosive and capable of dissolving most metals.

But of course discoveries such as these were only incidental to the quest for the Philosopher's Stone, which was often described by European alchemists in paradoxical ways. For example, according to the sixteenth-century work on alchemy, the *Gloria Mundi*, the Philosopher's Stone is

> . . . familiar to all men, both young and old, is found in the country, in the village, in the town, in all things created by God; yet it is despised by all. Rich and poor handle it every day. It is cast into the street by servant maids. Children play with it. Yet no one prizes it, though, next to the human soul, it is the most beautiful and precious thing upon earth, and has the power to pull down kings and princes. Nevertheless, it is esteemed the vilest and meanest of earthly things.

The *Gloria Mundi* seems to imply that anyone who found the Philosopher's Stone would surely fail to recognize it. Yet thousands of alchemists, some relatively unlearned and others with a vast knowledge of alchemical literature, continued to seek it. They pored over cryptic alchemical recipes and performed intricate experiments in their quest for the Stone, which they called the "Great Work."

Alchemical literature is almost always so cryptic, and contains so much obscure symbolism, that it borders on the unintelligible. For example, mercury was ordinarily not referred to by its common name. Instead it might be called doorkeeper, our balm, our honey, oil, May-dew, mother egg, green lion, bird of Hermes, or any of a large number of other names. Birds flying to heaven might represent distillation, and devouring lion might mean a strong acid. Copulation or marriage might represent certain alchemical procedures. A serpent or dragon could symbolize matter in an imperfect state.

There were probably many reasons for writing alchemical recipes in code. The Church frowned on the practice of alchemy so the practitioners must have wanted to maintain a certain amount of secrecy, which helped to avoid other dangers too. There was always the chance that some prince might demand that the alchemist produce gold, and then become very angry if the claimant couldn't.

Jabir had admonished:

> For heaven's sake do not let the facility of making gold lead you to divulge this proceeding or to show it to any of those around you, to your wife, or your cherished child, and still less to any other person. If you do not heed this advice you will repent when repentance is too late. If you divulge this work, the world will be corrupted, for gold would then be made as easily as glass is made for bazaars.

The message was clear: don't tell anyone how it might be done, or the gold you make might become worthless.

A LIFELONG QUEST

For some of the alchemists, the search for the Philosopher's Stone became a lifelong quest. One of the more extreme examples is Bernard of Treves, who sought the Stone from the time he was 14 until his death at the age of 85, squandering a fortune in the process. Bernard was born into a wealthy family in either Treves or Padua in 1406. As a child he often heard stories told by his grandfather about the alchemists' quest. Bernard became fascinated with the idea of seeking the Philosopher's Stone and began an intense study of the works of the Arabian alchemists. His family approved, having no objection to making the family fortune even greater.

The first book that Bernard discovered was al-Razi's *Secret of Secrets*. Setting up an alchemical laboratory, he spent four years and 800 crowns attempting to make gold. Unsuccessful, he turned to the works attributed to Jabir. By this time, news of what Bernard was doing had spread, and other alchemists flocked to him, offering their secrets and help. Neither their lore nor the writings of Jabir brought any success. But his assistants did succeed in parting him from a great deal of his money. After two years, Bernard found that he had spent another 2,000 crowns and was no nearer success.

When Bernard was 20, he met a Franciscan friar who told him stories about Pope John XXII, who had supposedly practiced alchemy, amassing a fortune of 18 million florins while issuing bulls against competition from other alchemists. Bernard and the friar studied the works of two well-known alchemists, Johannes de Rupecissa and Johannes de Sacrobosco, and decided that preparing highly distilled "spirit of wine" (alcohol) might help them achieve transmutation. They distilled the alcohol 30 times, until, as Bernard puts it, "it went off in such acridity that no glass could contain it." But again Bernard encountered only failure. The "Philosopher's Stone" that he created by this method did nothing.

Bernard next applied alchemical procedures to a vast number of different materials. He described his labors as follows:

> Twelve or fifteen years having been consumed in this manner and innumerable monies, without benefit, after the experiments of many received ones, in dissolving and congealing common, ammoniacal, pineal, saracen, and metallic salts, then more than a hundred times calcining them in the space of two years; also in alums of all kinds, in marcasites, blood, hair, urine, human dung and semen, animals and vegetables, in copperas, vitriols, soot, eggs, by separation of the elements in an Athanor by the alembic, and the Pelican, by circulation, boiling, reverberation, ascension, descension, fusion, ignition, elementation, rectification, evaporation, conjunction, elevation, subtilation, and commixtion: and other infinite regimens

of sophistications to which I stuck for twelve years having attained 38 years of age, still insisting upon extractions of the Mercuries from herbs and animals, thus had I uselessly dilapidated, as well by my own folly as by the seduction of imposters, about 6000 Crowns so that I became almost despondent. But nevertheless in my prayers I never forgot to beseech God that he would deign to assist my endeavors.

Bernard then encountered a magistrate of the city of Treves, who believed that the Philosopher's Stone could be obtained from sea salt. So Bernard set up an alchemical laboratory on the coast of the Baltic. He labored for a year and a half, working with the salt, but again encountered only failure, even though he repeated all of his procedures 5 or 10 times.

Bernard was now 48 years old, and he had been seeking the secret of transmutation for more than three decades. However, he was not one to give up. Because everything he had tried had ended in failure, he decided to travel through Italy, Germany, France, and Spain, seeking out alchemists wherever he went. During his travels, he encountered a monk named Gottfried Lepor who convinced him that eggs were an essential ingredient. So Bernard bought 2,000 hen's eggs, which he and Lepor boiled. They shelled the eggs and heated the shells in a gentle flame until they were perfectly white. Meanwhile they separated the whites from the yolks and putrefied them separately in horse dung. Next they distilled their materials 30 times, obtaining a white liquid and a red oil which, alas, were of no use in transmuting lead into gold.

At Berghem, in Flanders, an alchemist told Bernard that the Philosopher's Stone could be obtained from vinegar and copperas (green iron sulfate). So Bernard experimented with those materials too. Then, hearing that Master Henry, confessor to the Holy Roman Emperor Frederic III, had achieved success, Bernard set out for Vienna. When he arrived, he gave a lavish banquet, which most of the Viennese alchemists, including Master Henry, attended. Master

Henry confessed that he had not found the Philosopher's Stone but claimed that he did know of a method for increasing a quantity of gold. It was agreed that the alchemists present should contribute 42 gold marks. In five days, Master Henry said, these would be increased fivefold.

Master Henry began by making a paste of silver, mercury, and olive oil and placing it in a glass vessel. He held it over a fire and added the 42 marks before sealing the vessel and burying it in hot ashes around which a fire was kept up for 15 or 21 days, according to Bernard. After that time the vessel was broken and found to contain only 16 of the 42 gold marks. The other 26 had disappeared. Something caused the process to work in reverse and Bernard got back 4 of the 10 marks he had contributed, while the other alchemists had to share 12.

At this point, Bernard, now 58 years old, vowed to give up his alchemical quest. He kept the vow for two months, and then resumed his travels. He traveled to Rome, and then he went on to Messina, Cyprus, Greece, and Constantinople. He traveled to Egypt, Palestine, Persia, and England. According to Bernard, these journeys cost him more than 10,000 crowns, an enormous sum for those days. Finally, all his money exhausted, he returned to Treves.

Bernard's relatives in Treves considered him mad and would have nothing to do with him, so he decided to retire to the island of Rhodes and live there in modest circumstances. But then the inevitable happened. Bernard met a monk who was as enthusiastic as he was about learning the secret of transmutation. However, neither of them had the funds to buy the materials needed to carry out alchemical experiments. Bernard mortgaged what remained of his Italian estates, and once again he set to work. He worked with the same obsession as before and lived, slept, and ate in his laboratory. Even when he had exhausted the last of his money, he continued his quest, reading and rereading alchemical works.

Bernard lived in Rhodes until his death in 1490 at the age of 84, still trying to make gold. According to stories that grew up after his death, he did discover the secret of transmutation at the age of 81 and lived three more years, enjoying his newfound wealth. But of course Bernard himself made no such claim. In his writings, some of which were penned after he was supposed to have made this great discovery, there is no mention of having attained success. Bernard only warns seekers after alchemical truth that they should not allow themselves to be led astray by "imposters." Bernard's purported last words were, "To make gold, one must start with gold."

ALCHEMICAL FRAUDS

Master Henry, who conned Bernard and the other alchemists out of 26 gold crowns, was only one of the numerous frauds of those times. Unlike Master Henry, most claimed to have discovered the secret of transmutation and were prepared to demonstrate it. Their methods were many. A favorite was to use a knife or a nail made up of two halves that had been soldered together. The gold half was covered with a varnish that was soluble in alcohol, so that when the object was dipped in an alcohol solution, the varnish dissolved, making it appear that gold had been created. Other pseudoalchemists used double-bottomed crucibles in which gold filings had been concealed, or dropped pieces of charcoal in which gold leaf had been hidden into a crucible. Another common deception consisted of using an alloy of gold and mercury, so that when the substance was heated, the mercury was driven off and the gold left behind. The frauds then demanded large sums of money in return for their secrets or told their patrons that large expenses had to be met if they were to have any hope of producing gold in significant quantities.

Some of the fraudulent alchemists were quite flamboyant. For example, some time during the sixteenth century, an Arab came to Prague. He wore a turban and expensive flowing robes. On his arrival, he rented an expensive house and spent money lavishly. The Arab—if indeed he was an Arab—soon made the acquaintance of the many alchemists who lived in the city at the time. After he had been living in Prague for a time, he invited two dozen of them to a banquet and in the course of the evening announced that he could demonstrate a method for multiplying gold. Everyone who contributed a hundred marks, he promised, would receive a thousand when the procedure was completed.

After collecting the gold from his guests, the host took them to his laboratory, where he placed the coins in a crucible along with various alchemical preparations. He placed the crucible on a fire and seized a bellows with the intention of making the fire burn hotter. Suddenly there was an explosion, which filled the laboratory with scattered live coals, smoke, and noxious fumes. At the same time the laboratory was plunged into darkness.

Some of the guests soon found some candles and went back into the laboratory to see if their host was badly injured, but all they found was the broken alchemical apparatus and an open window. The Arab was gone. And of course the 2,400 marks had disappeared with him. This story might be somewhat embellished—it is the kind of tale that often is—but it certainly indicates that the con men of those days were not lacking in imagination.

Some of the pseudoalchemists succeeded in making off with large quantities of money. But others suffered less fortunate fates. In 1575 a woman named Marie Ziegler was roasted alive in an iron chair after failing to provide Duke Julius of Brunswick with a recipe for transmutation. In 1597 Georg Honnauer, who had promised to transmute iron into gold for the Prince of Wurtemberg, was caught putting gold into his crucibles. Honnauer was hanged on an iron gallows. One noble, Frederick of Wurtzburg, maintained a gilded gallows that was reserved for hanging alchemists who failed to keep their promises to make gold. On the gibbet there was the inscription, "I once knew how to fix mercury and now I am fixed myself."

In 1402 England passed an act of parliament that forbade the making of gold or silver by alchemical methods. The idea wasn't really to outlaw the practice but to give Henry IV, who was entitled to grant the right to make gold to certain people, a monopoly on gold making.

Henry hoped that alchemical gold might help him to pay state debts. In 1445 Sir Edmund Trafford and Thomas Asheton were duly granted the right to make gold, and coins were actually minted from the product they produced. But their alchemical "gold" later proved to be an alloy of mercury, copper, and gold.

Medieval literature has numerous references to pseudoalchemists and satires of alchemy. For example, around 1390 Chaucer satirized the alchemists in *The Canon Yeoman's Tale,* as did the English Renaissance poet John Lyly in his comedy *Gallathea* and Samuel Butler (the seventeenth-century English poet, not the nineteenth-century novelist of the same name) in *Hudibras.*

One of the best-known satires is *The Alchemist,* a comedy by Shakespeare's rival Ben Jonson, that targets not the pseudoalchemists but rather the gullible rich, who are so easily taken in. The play centers on the activities of Subtle, a butler who poses as an alchemist in his master's absence. With the aid of two accomplices, Face and Doll Common, Subtle swindles a number of people by engaging in quackery and he claims to be able to transmute gold. But of course comedies are supposed to end on a happy note. At the conclusion of the play Subtle's master returns unexpectedly, and his fraud is exposed.

The character of Subtle might be based on Simon Forman, who is mentioned by name in another of Jonson's plays. Forman, born in 1552, seems to have been a medical quack who sold love philters as a sideline. He was fined several times for pretending to cure the ill and was also sent to prison a number of times. In 1594 he began to tell fortunes and to experiment with transmutation. He attracted several wealthy customers, mostly women. Once he was asked to provide philters to the countess of Essex, who wanted to divorce her husband and win the love of the earl of Somerset. These facts came out during the murder trial of a woman who had acted as a go-between on behalf of the countess.

Alchemical frauds continued long after alchemy had fallen into disrepute. In 1867 three frauds bilked Emperor Franz Joseph of a sum equal to $10,000. In 1929 a plumber named Franz Tausend swindled a number of prominent German financiers after convincing them that he could make gold from lead. When Tausend was arrested, he claimed that his method was based on modern scientific ideas and asked for a chance to demonstrate the efficacy of his methods. He was taken to the State Mint where, in the presence of the director of the Mint, some police detectives, the state's attorney, and a judge, he produced a tenth of a gram of gold from one and two-thirds grams of lead. Because Tausend and all of his chemicals and apparatus had been searched before the demonstration, it appeared that the transmutation was genuine. However the following day it was discovered that gold had been smuggled to him in a cigarette while he was in jail.

THE DANGERS OF DECEPTION

In 1701 the 19-year-old Frederick Böttger, a German apothecary's apprentice, finding himself in need of money to continue his alchemical experiments, performed faked transmutations before some friends. If they gave him money to continue his quest, he told his onlookers, he would repay the money many times over. Böttger wasn't trying to defraud them, although that was how it turned out. He seems to have genuinely believed that he was on the verge of discovering the secret of the Philosopher's Stone. Indeed, he was so convinced that around this time he wrote to his mother assuring her that she would never lack for money again.

Böttger was aware of the dangers of performing such demonstrations. More than one avaricious prince had meted out severe punishment to alchemists who claimed to be able to produce gold and then failed, so the alchemist pledged the witnesses of his transmutations to secrecy. This didn't prevent rumors from spreading, but rumors of this sort were common in those days, and in any case they didn't spread widely enough to get Böttger into trouble. But then he made a big mistake. In October 1701 his employer, the Berlin apothecary

Frederick Zorn, released Böttger from his apprenticeship. He was now a journeyman who could work for wages. During the years of his apprenticeship Zorn had always been critical of his alchemical experiments. Alchemists had been searching for the Philosopher's Stone without success for centuries, Zorn said, and added that Böttger would do better to master the preparation of medicines than pursue some hopeless quest. So when Böttger left Zorn's employ he staged yet another demonstration for Zorn and two of the apothecary's friends, melting down some silver coins and turning them into gold. One might think that Zorn would guess that the demonstration was fraudulent. If Böttger really could make gold, he would not have had to continue apprenticeship. But apparently this didn't occur to Zorn and his friends. They were convinced that the transmutation was real. And in spite of Böttger's request that they remain silent, they began to talk. Their talk aroused great interest in Böttger's feat. After all, Zorn was no impulsive youth spouting tales of another youth's demonstration. He was a leading Berlin apothecary, and his words carried weight.

It didn't take long before the Prussian king, Frederick I, heard about what happened in the apothecary's shop. Frederick immediately summoned Zorn and questioned him about the transmutation that he had witnessed. Frederick seems to have been impressed by Zorn's account because he ordered the apothecary to return the next day with his former apprentice. Meanwhile he confiscated the gold that Böttger had supposedly manufactured. When Böttger heard of the interview, he realized that this wasn't good news and immediately went into hiding. But Frederick was not to be denied. When Böttger failed to appear at his court, he offered a substantial reward for the alchemist's capture. This was enough to convince Böttger that his only recourse was to slip out of the country. Fortunately, he was able to persuade an acquaintance to hide him in a covered wagon that was being driven to nearby Saxony.

Böttger enrolled as a medical student at the University of Wittenberg. But Frederick soon discovered his whereabouts and sent a detachment of troops to capture the young fugitive. But Frederick couldn't just take Böttger back to Prussia without the consent of the Saxon authorities. Doing so would damage relations with Augustus, the elector of Saxony. Besides, by now the Wittenberg authorities weren't anxious to let Böttger go. The tales of his gold making had spread, and it did little good for the Prussians to insist that the fugitive was just a common criminal. The Wittenberg authorities sent to the elector, asking for instructions about handling the affair. They were not answered immediately because Augustus, who was also king of Poland, was then in Warsaw. Weeks passed, during which nothing happened. The Prussians continued to demand that Böttger be given into their custody, and the Saxon officials continued to delay.

Finally a message from Augustus arrived. He ordered Böttger imprisoned in Dresden, the capital city, until he revealed his method for making gold. The Saxons knew that the Prussian soldiers might become desperate and use force to seize the prisoner while he was en route from Wittenberg to Dresden so they provided Böttger with a military escort. Meanwhile, in order to preserve the illusion that he was still in Wittenberg, Saxon soldiers continued to stand guard outside Böttger's lodgings, and food was carried in for the next two days.

IMPRISONMENT IN DRESDEN

In Dresden Böttger was confined in a section of the royal castle that was equipped with a laboratory. He was given three assistants to help him pursue his quest for gold, and two members of Augustus's court were assigned to supervise the work. Böttger was allowed to talk to no one other than these five. But of course they were not his only human contacts; he also had his guards.

Augustus was impatient to witness a transmutation. He ordered his prisoner to send a sample of the Philosopher's Stone to Warsaw as soon as he could. This created a dilemma. Böttger could hardly admit that he didn't know how to make gold. If he did and he was not believed, there was a good chance that he might be tortured for the recipe he didn't have. So he sent a box containing some alchemical apparatus and some mercury and other ingredients to Augustus, along with instructions for making a small quantity of gold. Böttger's instructions were followed in an experiment performed in Augustus's Warsaw palace, but all that was produced was a metallic mass that looked nothing like gold. But this didn't discourage Augustus, who commanded that the alchemist continue to be confined while he carried out further experiments.

Augustus also ordered that Böttger be allowed a more comfortable imprisonment. The alchemist was given two rooms in Augustus's Dresden palace and allowed contact with people other than his assistants and jailers. Böttger responded by making grandiose promises to the king. He claimed, for example, that he would soon be producing large quantities of gold every month. He must have quickly regretted these promises and, fearful of the consequences if Augustus discovered he hadn't made any gold at all, he determined to escape. Because he was lightly guarded, he didn't find it difficult to slip away from the palace and make his way to a meeting place where a friend waited with a horse. He rode into Austria and then headed toward Prague. But his freedom didn't last long. A party of Augustus's soldiers traced him to an inn in the town of Enns, where he had stopped to rest, took him into custody and brought him back to Dresden. He enjoyed only five days of freedom.

The greedy king still believed that Böttger would eventually find a way to produce gold. After consulting with members of his court, he decided to spare the young man harsh punishment but kept him under closer guard than before. But Augustus was not a man of limitless patience. In 1705, when Böttger had been a prisoner for more than three years, Augustus demanded that his prisoner set a definite date by which gold would be produced. Böttger again made grandiose claims. He wrote a document promising to produce gold within 16 weeks and to manufacture 2 tons of the precious metal during the following 8 days.

When Böttger failed to fulfill his promises, the king was furious and was inclined to have his prisoner summarily executed. However his advisers pointed out that doing so would cast doubts on Augustus's judgment. After all, he had spent large sums of money over a period of years financing Böttger's experiments. One of these advisers was Ehrenfried Walter von Tschirnhaus, whom the king employed to find new mineral deposits and inaugurate new manufacturing projects. One of Tschirnhaus's pet projects was finding a way to manufacture porcelain. Böttger would be an excellent candidate to continue the project when Tschirnhaus grew too old to continue the quest himself. Not only was Böttger a brilliant chemist, he was also still a young man. Tschirnhaus, on the other hand, was growing old. Augustus listened. He was an enthusiastic collector of porcelain himself.

WHITE GOLD

From the time that they first appeared in Europe during the sixteenth century, Chinese porcelain objets d'art were highly prized. Porcelain was far harder than any other ceramic material, and it exhibited a translucence that no European pottery could match. The first porcelain pieces to arrive in Europe inevitably found their way into the treasuries of European rulers. Then, as the porcelain trade grew, wealthy aristocrats began collecting objects made of the precious material. Europeans potters naturally looked for ways to manufacture porcelain themselves. If they discovered the secret, the profits would be immense. However, the secret of manufacturing porcelain turned out to be as elusive as the secret of the Philosopher's Stone.

The translucence of porcelain suggested to most European potters that the material must be a combination of clay and glass, and they tried using many different combinations of glass, clay, and other materials. Some of them succeeded in producing materials that bore a superficial resemblance to porcelain, but the pottery lacked the fineness of Chinese porcelain. Their mistake was assuming that glass was an ingredient. It wasn't. Chinese porcelain was made by mixing white clay with a pulverized stone that contained feldspar and then firing objects fashioned from these materials at high temperatures. During firing the two materials fused together, producing a hard, flawless, non-porous material.

When Tschirnhaus suggested that Böttger be put to work on the porcelain project, the king listened with great interest. He quickly decided that there was no reason why Böttger couldn't find a way of making porcelain while continuing his alchemical experiments. He had Böttger transported to the Albrechtsburg, a royal castle at Meissen, 9 miles from Dresden, which had ample space to set up a larger laboratory. Böttger's life at Meissen was not as comfortable as the one he enjoyed in Dresden. The Albrectsburg had been unused for some time and had been pillaged during the Thirty Years War. However, Augustus cared much less about his prisoner's comfort than about the money that a successful porcelain factory might earn.

Böttger was provided with five assistants, and 24 furnaces were built in the laboratory. Samples of clay from all parts of the kingdom were sent to him. The windows of the castle were bricked up so that passersby could gain no inkling of what was going on, and Augustus ordered Böttger and his assistants not to discuss their work with any- one but the courtiers that Augustus appointed to supervise them.

Böttger made no attempt to produce porcelain by mixing clay and glass together. He knew that this had been tried many times and had led only to failure. Instead, he began a series of careful experi- ments in which he mixed various different clays with different kinds of rock and fired them at high temperatures. Only at high tempera- tures, he realized, would the rock was be melted. Within a year, Böttger had achieved some success. He hadn't produced porcelain, but he had learned how to make a red stoneware that was finer than anything produced by other German potters. It wasn't white, and it wasn't translucent, but it was a new ceramic material. He had every reason to expect further progress.

But then his work was suddenly interrupted. As king of Poland, Augustus was at war with Sweden. He had suffered a disastrous defeat, and Swedish troops were now advancing toward Dresden. Augustus ordered his most valuable possessions transported to Königstein, an impregnable country fortress that stood on a rocky plateau above the Elbe River and was used as a prison. And of course these valuable possessions included Böttger. Augustus couldn't let his prize alchemist fall into enemy hands.

Böttger remained at Königstein for the next year. It was a time of inactivity. There was no laboratory in the castle and few ways to pass the time. In effect, he had been transferred to the eighteenth-century equivalent of a maximum-security prison. At first Böttger was not allowed to have even books, ink, or paper. When, after some time, he was given writing materials, he wrote a series of despairing letters to the king, pleading for a chance to continue his work. The political situation was too unsettled for Augustus to allow that. If Böttger was allowed to leave Königstein, he could easily fall into the hands of the Swedes. But in 1707, Augustus abdicated as king of Poland and the Swedish forces withdrew from Saxony. Augustus soon ordered that a new laboratory be set up in Dresden, and when it was finished, Böttger was allowed to leave Königstein. However Augustus was in no mood to be told of any more failures. He informed Böttger that, if he failed to produce either gold or porcelain, he would be executed.

Böttger apparently decided that he had a better chance of making porcelain than of transmuting base metals into gold in the foreseeable future. He resumed his series of experiments with different combinations of clay and minerals. A material he thought especially promising was China clay, a mineral that was mined in Germany but also found in China. China clay contains feldspar, but Böttger achieved nothing with it because the temperatures that his furnaces could achieve were too low to melt the material. So he decided to try alabaster, a type of gypsum that is snow white and translucent. Böttger mixed clay and alabaster together in different ratios and found that if he used seven to nine parts of clay to one of alabaster, a hard, white, translucent material was produced. He had succeeded in making porcelain! There was, of course, much more to be done.

Böttger's first porcelain pieces were not of the same high quality as those imported from China, and he did not yet know how to produce a glaze. However he was confident that he could perfect his techniques and eventually produce porcelain of finer quality.

Augustus was pleased but he kept Böttger confined, still expecting him to find a way to make gold and intending to keep him imprisoned until he did. The alchemist was still kept in Dresden, even after a porcelain factory was set up at Meissen and he was appointed its director.

Augustus made Böttger a baron in 1711, and thereafter the alchemist lived the life of an aristocratic gentleman. Nevertheless, his imprisonment continued. The king (Augustus had regained the Polish crown in 1710) had no intention of letting him go free before he found the Philosopher's Stone. Augustus relented only when Böttger became very ill in 1714. Although he was only 32, his eyesight was failing, and he began to suffer epileptic seizures and a consumptive fever. Böttger's illness probably had a number of contributing causes. Ever since Augustus first imprisoned him, he had been a very heavy drinker. It was said that during the latter part of his life he rarely spent a day sober. And of course anyone who labored in an alchemical laboratory was likely to inhale poisonous fumes, especially from arsenic and mercury, which were commonly used in alchemical experiments at the time.

Once he was free, Böttger's health seemed to improve, but it soon became obvious that this was an illusion. During the next few years he became increasingly weak and died in 1719 at the age of 37. He had spent more than 12 years as a prisoner and had been free only during the last 5 years of his short life.

THE NEVER-ENDING QUEST

Alchemy was supposedly superseded by chemistry in the eighteenth century. But alchemical practices never really died out, and today there are still people who persist in practicing the art. There are publishers and book dealers who specialize in alchemical books, and there are alchemical groups and societies. Alchemical elixirs and tinctures can be purchased on the Internet, and one can study alchemy at Paracelsus College in Australia.

Modern alchemists do not always attempt to make gold. The mystical alchemy that I previously spoke of still exists today. It is called esoteric alchemy, and it often becomes intermixed with other "new age" and mystical ideas. For example, the various Rosicrucian groups make use of alchemical concepts and mysticism, and herbal remedies are sometimes said to be made by alchemical methods. Sometimes alchemy is seen as nothing more than a path to spiritual growth. For example, one website that I consulted informed me that "the main goal of Alchemy is the creation of a spiritually complete individual whose several components of consciousness are united, resulting in an integrated, independent, enlightened human being." In other words, it is spiritual gold that is sought. However, you shouldn't imagine that modern alchemists have given up their quest to make gold. In fact you can find recipes for doing precisely that at the following website: www.dnai.com/~zap/gold.htm.

CHAPTER 2

Prelude to the Birth of Chemistry

During his lifetime and long after his death, numerous stories and legends circulated about the Swiss-German physician and alchemist Paracelsus. In the eyes of many, Paracelsus was more than a doctor. He was a magician and soothsayer who had learned the secret of immortality and could resurrect the dead. It was said that he had been seen in several different places at the same time and that he rode a white horse given to him by the Devil. It was said that he could change brass coins into gold and that he spoke with spirits. More than 300 years after Paracelsus's death in 1541, people made pilgrimages to his grave in Salzburg during a cholera epidemic, hoping to be healed by occult powers still at his command.

It wasn't just the poor and ignorant who believed the tales about Paracelsus. The scholarly English poet and cleric John Donne accused Paracelsus of serving the Devil. In time the stories about him became intertwined with those about the legendary Dr. Faustus. And when Goethe wrote *Faust* two and a half centuries after Paracelsus's death, he incorporated several allusions to the magician's life and writings.

But Paracelsus wasn't only a healer and an alchemist, he was also a scientist. Declaring that alchemists were wasting their time trying to make gold, he engaged in chemical experimentation. Wanting to find new remedies for different kinds of ailments, Paracelsus studied chemical reactions and investigated the properties of chemical substances. He believed that illness was caused by chemical imbalances within the body and, therefore, it was important to study nature in order to find medicines that could correct these imbalances. Paracelsus was the first to use the word "chemistry." He proposed that chemical compounds be prepared from pure chemicals, and that when they were used as medicines, they should be administered in precise doses. These were new ideas. The alchemists of the day ordinarily gave little thought to the purity of the materials they worked with, and the physicians of the time were generally very careless about the quantities of the remedies that they prescribed. Furthermore, Paracelsus attempted to classify chemical substances according to the kinds of reactions they produced. Thus he distinguished between magisteria, specifics, elixirs, quintessences, tinctures, and mysteria, and while none of these classifications would mean much to a modern chemist, the idea that chemicals should be assigned to different groups was a great step forward.

Although he was widely regarded as a magician who could effect miraculous cures, Paracelsus was no charlatan and he made some real contributions to medicine. He believed in the healing power of the human body and railed against the methods then used to treat wounds, such as applying cow dung, viper fat, or feathers. "If you prevent infection," he said, "Nature will heal the wound all by herself." He published the best clinical description of syphilis that had been written up to that time and advocated treating the disease with limited doses of mercury. This remained the standard treatment until the discovery of salversan in 1909. Paracelsus was not the first to use mercury to treat the disease. However, he clearly realized that it was far more effective than the other remedies then in use, and he avoided giving his patients large quantities of the poisonous metal.

He studied silicosis ("miner's disease") and concluded that it was the result of inhaling vapors in mines, not a kind of revenge inflicted by mountain spirits. He was the first to realize that goiter could be caused by minerals in drinking water, and he prepared and used a variety of new chemical remedies for the condition. In effect, he invented chemotherapy.

THE BOMBASTS

Paracelsus's real name was Philipus Aureolus Theophrastus Bombastus von Hohenheim. "Paracelsus" was a name he gave himself as an adult. It meant "greater than Celsus." Paracelsus regarded himself as superior to Celsus, a famed first-century Roman physician whose writings had recently been rediscovered. "Aureolus" was Paracelsus's alchemical name and was also one that he gave himself. Aureolus was an ancient alchemist whose writings were in the library of Paracelsus's father.

Paracelsus was born in 1493, the year Columbus returned from his first voyage to the New World. The Bombasts ("Bombastus" is the latinized form of Bombast) were an aristocratic German family. Originally known as the Banbasts von Hohenhain, the name later became corrupted into Bombasts von Hohenheim. Paracelsus's grandfather, Ritter Georg von Hohenheim, was a commander of the Teutonic Knights who became known for an adventurous pilgrimage to the Holy Land in 1468. After he returned to Germany, he became involved in a political feud and fell into disgrace for speaking intemperately and impudently during official sessions of the high court of justice in Stuttgart. He later lost his estate, which was sold at auction.

Paracelsus's father, Wilhelm von Hohenheim, was one of Georg's illegitimate offspring and thus had no legal right to use the noble "von" as part of his name. However, he apparently didn't worry very much about legal niceties; he went ahead and used it anyway. Wilhelm seems to have had an impoverished youth. The students' register of the University of Tübingen listed him as a "pauper." Nevertheless,

Wilhelm managed to earn a degree of licentiate in medical sciences from the university. However, lack of funds might have prevented him from pursuing his medical studies further because he didn't attempt to earn the title of "Doctor of Physick." Instead, he wandered through southern Germany and then into Switzerland, finally settling down as a country doctor at an inn next to the Devil's Bridge, which crossed the Siehl River. In 1492, Wilhelm married Elsa Ochsner, a bondwoman at the nearby Benedictine Abbey of Einsieden. On St. Philip's Day of the following year, Elsa gave birth to a son, who was named Philip, after the saint. At the christening, he was named Theophrastus, after Aristotle's successor at the Lyceum.

Theophrastus was frail and sickly as an infant and suffered from rickets as a child. The effects of the disease can be seen in an etching that was made of Paracelsus as an adult. According to legend, he was also a eunuch, the result of an encounter with a wild boar. Another story says that he was castrated by some drunken soldiers. No one really knows which story is true or whether Paracelsus was indeed emasculated. Portraits of him as an adult never show him as having a beard, and he never became involved with any woman. Furthermore, he once declared that it was better to be a eunuch than an adulterer. On the other hand, he became bald in his later life and eunuchs almost never go bald.

Paracelsus's mother, Elsa, was apparently a manic-depressive. When he was nine years old, she committed suicide by jumping from the Devil's Bridge into the Siehl River. After her death, Wilhelm von Hohenheim left the isolated valley and settled with his son in the German city of Villach. Wilhelm, who knew a great deal about mineralogy, soon found employment at a nearby school of mining operated by the Fuggers, the great bankers of the day.

THE WANDERING SCHOLAR

In 1507, at the age of 14, Paracelsus became a wandering scholar, traveling from university to university in search of knowledge. Such a life wasn't uncommon in those days. Students often went from school to school, seeking out famous teachers or the kind of instruction they desired. The scholars generally traveled in groups, making their living by doing such things as singing in inns, pulling teeth, selling drugs, or begging.

For two years, Paracelsus wandered through Germany, never staying very long in one place. He went to the universities at Heidelberg, Leipzig, Wittenburg, and Cologne and to his father's alma mater at Tübingen. Next he went on to Paris and Vienna and then to Italy. It is unclear whether Paracelsus got the baccalaureate in medicine from the University of Vienna in 1510. In later life he sometimes said that he got the degree from an Italian university, without specifying precisely where. He also claimed to have received a doctorate from the University of Ferrara in 1516. However, university records for that year are missing, and it is impossible to verify his claim. There is reason to doubt that he really was awarded the degree; he was never able to produce any proof that he received it.

If the details of Paracelsus's education are uncertain, it is clear that he did not have a high opinion of the academic teaching of his day. At the age of 19, he said, "At all the German schools you cannot learn as much as at the Frankfurt Fair." And he later wrote: "All the universities and all the ancient writers put together have less talent than my ass." He was especially critical of medical education, which was based on the teachings of the ancient Greek physician Galen and on such Arab medical authorities as the eleventh-century philosopher and physician Avicenna. Neither did he have a high opinion of those who earned the doctoral degree. Speaking of contemporary physicians, he said, "Their ignorance cannot justify their fantastic theories. All they can do is gaze at piss." (At the time, urine inspection was the most commonly used method of diagnosis.)

YEARS OF WANDERING

After his university education ended, Paracelsus spent years wandering through almost every country in Europe seeking out knowledge.

If he couldn't acquire the medical knowledge he sought from the universities, he would find it elsewhere. "A doctor," he wrote, "must seek out old wives, gypsies, sorcerers, wandering tribes, old robbers, and such outlaws and take lessons from them."

Scholars have been unable to reconstruct all the details of Paracelsus's wanderings between 1517 and 1523. It is known that he traveled thousands of miles, but it is uncertain how long he remained in any place or how he made a living. By his own account, he traveled through Spain, Portugal, England, Brandenburg, Prussia, Lithuania, Poland, Hungary, Transylvania, Croatia, and other lands. Traveling west, he visited four great universities, those of Montpellier in southern France, Seville and Salamanca in Spain, and the Sorbonne in Paris. He was not very impressed by the scholars at any of these schools, however. He wrote that at the Sorbonne "the Parisian doctors despise all others; yet they are ignoramuses. They think their high necks and high judgment reach into heaven." When he traveled to England, he neglected to visit either Oxford or Cambridge.

After returning to the continent, Paracelsus joined an insurgent Dutch army as a surgeon. Two Dutch provinces had revolted against their former ruler, Charles I of Spain. When the rebels were beaten, Paracelsus moved on to Denmark, where he is said to have cured King Christian's mother of an illness. In 1520 the Danes invaded Sweden, and Paracelsus accompanied the army as a master surgeon. Although Christian was initially successful in subduing the Swedes, they soon shook off Danish domination, and Paracelsus moved on again. He found work as a surgeon with the Teutonic Knights but didn't stay long. According to one story, when Basil, the Russian grand duke invited Western physicians, astrologers, architects, and humanists to his court in Moscow, Paracelsus accepted with alacrity.

Paracelsus didn't remain in Moscow for long, if indeed he went there. Shortly after he arrived, the Tatars invaded the new Russian state, sacked Moscow, and took the foreign scholars to their capital. The tales about Paracelsus tell us that he was treated quite well by his captors. The Tatars considered healers to be holy men. In 1521 he supposedly accompanied a Tatar prince on a diplomatic mission to Constantinople, which was then ruled by the Turks, and was thus able to gain some knowledge of Byzantine alchemy. He later claimed to have gone on to travel Egypt, Persia, and the Holy Land. In fact, he never visited these places. When speaking of his travels, he was apparently not afraid to embellish the truth.

Contemporary accounts of Paracelsus's behavior during his travels are contradictory. According to one, "He lived like a pig and looked like a sheep drover. He found his greatest pleasure amongst the company of the most dissolute rabble, and spent most of his time drunk." But other contemporaries described him in such terms as "the German Hermes" or "the king of all knowledge." There may have been a great deal of truth in both kinds of description. After Paracelsus's death, his secretary, Oporinus, wrote, "The two years I passed in his company he spent in drinking and gluttony, day and night." But according to Oporinus, "Noblemen, peasants and their womenfolk adulated him like a second Asclepius [the Greek god of healing]." Oporinus might not have been exaggerating, because Paracelsus treated rich and poor alike, asking high fees from the well-off while tending to the poor for nothing.

TOWN PHYSICIAN

In 1524 Paracelsus returned home and lived for a while with his father, Wilhelm. He brought back with him a large sword, which he claimed to have gotten while serving in the Venetian army and which he kept always at his side, even when he slept. He also had a supply of a drug called laudanum, which was probably an opiate. He was said to keep the laudanum in the hollow pommel of the sword. This sounds likely, because he never used the sword as a weapon. He also brought back the knowledge that became the basis of his doctrines in medicine and chemistry. He had traveled through a large part of the known world and had seen the diseases that were common in different places. He had seen healers who were ignorant of Galen work their cures. He had made contact with numerous alchemists and had acquired a wealth of information about chemistry, myths, folklore, and methods of healing.

Wilhelm was happy to see his son, but Paracelsus didn't stay long. Unwilling to practice medicine in Villach in competition with his father, he went to nearby Salzburg. However, he wasn't there long before he became involved in a peasant uprising. The extent of his involvement is uncertain, though he fled Salzburg when the insurrection was suppressed. He went next to Strasbourg. As a "doctor of both medicines," he joined the Guild of the Alfalfa, which had a membership consisting of doctors, millers, and grain merchants. Strasbourg was the one city in Germany where surgeons were equal in status to physicians. In Paracelsus's day barbers still performed many minor operations and surgeons were generally not more highly thought of than barbers. A physician had the right to ride on horseback and to wear a fur hat and a red robe. Furthermore, he had the right to marry into the aristocracy. Surgeons, on the other hand, were plebeians. In most places physicians were not supposed to associate with surgeons, and they certainly would never contemplate performing surgery.

At the time, Strasbourg had the best surgery school in Germany. It was directed by two celebrated surgeons, Hieronymus Brunswick and Hans von Gerstorff. When Paracelsus came to Strasbourg, he fully expected to be allowed to teach at the school, but his hopes were dashed when one day a member of the faculty, Dr. Vandelinus Hock, challenged him to engage in a public discussion of anatomy, a subject about which he knew little. Unable to answer Hock's questions, Paracelsus quietly left the lecture room.

However, this setback did little to diminish his reputation as a physician. In 1527 he was called to Basel, a city some 70 miles away. The wealthy publisher and humanist Johan Froben was ill with an infected leg. The local doctors had recommended amputation, a very dangerous procedure. In those days, many people died after having a limb amputated. Paracelsus moved into Froben's house and was able to cure him without resorting to such drastic measures. At the time the great scholar, Erasmus of Rotterdam, was staying with Froben. Erasmus asked Paracelsus to treat him for gout and a kidney ailment, and the healer again came up with successful cures. Paracelsus's reputation soared, and on the recommendation of Froben and Erasmus, the town council appointed him municipal physician and professor of medicine at the University of Basel.

The physicians on the university faculty were not happy to see Paracelsus made a professor without their consent. So they insisted that a colloquy, a public test of the new professor's qualifications, be held. Under ordinary circumstances, a colloquy was only a formality but not this time, as Paracelsus realized when he discovered that Dr. Hock had come from Strasbourg to debate him. In the end, Paracelsus simply did not show up for the colloquy. He and his allies on the town council contended that because the council had created the position, the appointment did not have to be validated by the university. The university countered that because it had not been allowed to confirm Paracelsus's appointment, it did not have to extend any academic privileges to him.

When the summer term came near, Paracelsus posted an announcement of his lectures, declaring that he would lecture in German, rather than in the traditional Latin. For two hours every day he would explain his healing methods. He also promised not to rely on such ancient authorities as Galen, but to teach methods that were based on experience. "What a doctor needs," he wrote, "is not eloquence or knowledge of language and books, but a profound knowledge of Nature and her works." Another innovation was that barbers, alchemists, and others lacking academic background were to be admitted to the lectures, along with registered students. Naturally this did not go over well with the faculty, and the dean of the university announced that if the university had no jurisdiction over Paracelsus or the content of his lectures, it need not make a lecture room available. Paracelsus was forced to find an off-campus hall for his lectures.

Thirty students (22 more than had registered at the university the preceding year) attended the opening lecture. The room was also filled with townspeople, academics, and physicians. The physicians apparently attended in order to see what nonsense this fool would drum up. Paracelsus made a dramatic entrance. He walked to the lectern wearing a professor's robe, which he then threw out of the room, revealing a dirty alchemist's apron. He then announced that he would now reveal medicine's greatest secret—and dramatically uncovered a dish of excrement. When the physicians began leaving the hall, Paracelsus taunted them, saying, "If you will not hear the mysteries of putrefactive fermentation, you are unworthy of the name of physicians."

Paracelsus went on to dismiss the theory on which the orthodox medicine of the day was based. This theory, which had originally been proposed by Hippocrates, held that the body contained four humors: blood, phlegm, yellow bile, and black bile. Disease was supposedly a consequence of imbalances in these humors, and it was the physician's job to correct the imbalances. Furthermore, each humor was associated with one of the four elements. For example, a fever was clearly the result of the presence of too much fire. The humor that corresponded to fire was blood, so feverish patients should be bled. All of this was nonsense, Paracelsus said. The body was a kind of chemical laboratory, and a doctor must investigate the properties of chemical compounds to find those that would cure any specific disease.

All of this sounds very levelheaded. However, it would be a mistake to view Paracelsus as a medieval version of a modern doctor. On the contrary, he was quite capable of inventing fantastic theories, and indeed he did. According to his doctrine, plants and minerals had "signatures" that a wise man could learn to read. For example, orchids were shaped like testicles, indicating that their juice would "restitute his lewdness to a man." Similarly, black hellbore, which bloomed in winter, had the power of rejuvenation, and liverwort and kidneywort had the shape of the parts of the body that they could be used to cure.

If some of what Paracelsus taught was nonsense, his students gained a better medical education than they would have from professors lecturing from the medical classics. Rather than blindly follow theory, Paracelsus had spent years seeking out practical knowledge about healing and he had a great deal to impart. And he was certainly popular with the students. After Paracelsus left Basel, enrollment at the university dropped to *one* undergraduate.

CONTROVERSY AND DEPARTURE

Paracelsus's lectures had incited controversy and he didn't shy away from provoking more. June 24 was St. John's Day, a day of commencement celebrations when students traditionally hurled whatever they no longer wanted or needed into a bonfire. It was also only three weeks after Paracelsus's inaugural lecture. Cheered on by boisterous students, he joined the merrymaking, publicly burning a book by Galen and the *Canon* of Avicenna.

Having already outraged the physicians and academics, Paracelsus quickly proceeded to make even more enemies. As municipal doctor, he made full use of his authority to inspect the local pharmacies. On a tour of inspection of the local shops he condemned the preparations that filled the pharmacists' shelves as worthless and branded their prescriptions as "foul broths." Naturally the pharmacists' business suffered as a result. Meanwhile, he antagonized the physicians further by attempting to supervise their prescriptions, at the same time roundly criticizing their methods of treatment.

A response was not long in coming. His opponents had gained support among some of his students and were kept informed of everything that was said at his lectures. Then one day a lampoon of Paracelsus was found nailed to the church door. Written in the form of a poem, it was purported to have been written by the shade of Galen speaking from Hades. It referred to Paracelsus as "Theophrastus, or rather Cacophrastus [the name can be translated from German as "Crapophrastus"]," and charged that he was "not worthy of carrying Hippocrates' piss-pot." Paracelsus did not take this at all well. He immediately sent a letter to the town council, asking it to find the perpetrators and punish them. But the council had no desire to punish student pranks, and did nothing. Meanwhile, Paracelsus continued to assail the local doctors and pharmacists, calling them cheats, frauds, and "ass-scratchers."

He was rapidly becoming an embarrassment to the town council, but he still had the support of the influential Froben and thus was able to retain his position as municipal physician. But not for long. While his protégé was away in Zurich, the aging Froben rode 400 miles on horseback to the Frankfurt Fair, against Paracelsus's explicit advice, and died there of a stroke. Not only did this eliminate Paracelsus's strongest supporter in Basel, but it was a potential source of trouble, because his enemies immediately began spreading the story that Froben had been poisoned by his doctor's chemical medicines. The university faculty persuaded the town council to order an investigation into the causes of Froben's death. Because doctors could be held responsible for the health of their patients, Paracelsus was in danger of being prosecuted.

The investigation was never concluded because Paracelsus managed to quickly bring about his own downfall. A wealthy canon called on the controversial physician to treat him and offered a fee of a hundred guilders if he succeeded. This was quite a large amount of money considering that a court physician might earn a hundred guilders in a year. Paracelsus duly cured the canon in a short time and demanded his fee. But the canon now offered only six guilders, saying that that should be a sufficient amount for so little work. Indignant, Paracelsus sued the canon for the remainder of his fee. But when the court ruled in favor of the canon, Paracelsus became even more enraged. He made the same kind of mistake that his grandfather had once made. He penned a lampoon, distributed as an anonymous pamphlet, in which he denounced the magistrates as corrupt. Libeling the magistrates was a serious offense, and of course everyone knew who had written the pamphlet. In order to avoid arrest, Paracelsus was forced to flee Basel in the middle of the night.

FURTHER WANDERINGS

He fled to Colmar in Alsace, where he accepted the hospitality of his friend, Dr. Lorenz Fries. However, the Colmar town authorities had heard of what had taken place in Basel and allowed Paracelsus to stay only temporarily, so he went to Esslingen, a small town in Württemberg. He hoped to settle there, but was soon forced to leave. It isn't clear what kind of trouble he got into. He later wrote, "My misfortune, which began at Esslingen, was confirmed at Nuremberg." However, he didn't give details.

The citizens of Nuremberg had also heard of him. "They say he burned Avicenna at the University of Basel, and he stands alone against nearly the whole medical guild," one of them wrote. On Paracelsus's arrival, the local physicians challenged him to a debate. He declined and asked that instead he be given a patient they considered incurable, preferably one with syphilis, so that he could prove that he could bring about cures. It happened that 14 or 15 patients were quarantined in a leper hospital outside Nuremberg. "Leprosy" was the common name for a number of different skin diseases among which the physicians of the day did not distinguish. Paracelsus succeeded in curing 9 of the patients. No case histories survive, but it is likely that those 9 were syphilitics and that Paracelsus temporarily relieved their symptoms.

But when this feat did not bring him the acclaim he desired, Paracelsus denounced the entire local medical establishment. He railed against their opulence, accused them of cheating the public, and spoke of their "buxom fat wives." And as though that were not enough to get him into trouble, he then inveighed against the local Lutherans, criticizing their doctrines. And *then* to cap it all off, Paracelsus precipitated a conflict with the powerful Fugger family. At the time, guaiac, a bush that grows in the West Indies, was commonly used to treat syphilis. The physicians put their patients in closed rooms in which the guaiac was burned. Though this did nothing to relieve the symptoms of syphilis, the use of guaiac became quite fashionable within the medical establishment. Realizing how profitable the importation of guaiac could be, the Fuggers obtained a monopoly.

Paracelsus realized that the treatment was worthless and published a pamphlet against the use of guaiac. This pamphlet was to have been followed by a book, titled *Essay on the French Disease*, in which the use of mercury treatments was recommended. But only the first chapter of the book was printed, because the censor forbade the printing of the remaining two chapters. It wasn't difficult to figure that the Fuggers had a hand in the censorship. They couldn't allow an itinerant doctor to threaten their profitable business. However, Paracelsus persuaded another publisher to issue a clandestine edition, and *Three Chapters on the French Disease* appeared in 1530. To add to the insult, Paracelsus dedicated the book to Lazarus Spengler, the secretary of the aldermen. The reaction was immediate. Again, Paracelsus had to flee.

PREACHING POVERTY

For the next several years Paracelsus wandered through Switzerland and eastern Germany, never staying anywhere very long, not always by choice. For example, a 1532 police record states that he was "driven out" of Prussia. He went from one place to another, making a living however he could, and tending to the ill. Describing this period, he wrote, "I do not know where to wander now. I do not care either, as long as I help the sick." Weary and destitute, he returned to Switzerland, where he had friends among the peasants and teamsters. They were the company he preferred, not just because he liked to drink with them. "For I have found that in the common man and peasant," he said, "Christian life is most perfect."

Wealth and fame eluded Paracelsus. Perhaps that is why he now turned his thoughts to religion. He no longer railed against orthodox medicine or proclaimed the superiority of his own methods. Instead, he thought of God, salvation, and the Christian life. During this period, he preached to the poor, telling them that, "Blessed and more blessed is the man to whom God gives the grace of poverty. . . . You should therefore sell everything and give [the proceeds] to the poor." He railed against luxury, going so far as to condemn even the spice merchants:

> They are pepper-traders and deal in spices. What good is it? Whom does it benefit? . . . [his ellipses] Does not a peasant live as well on turnips as one who eats spices? All traders in spices are full of the devil and his servants, through whom he disgraces the people.

Paracelsus fasted from time to time and gave away his clothes and what money he had. He even abstained for a time from chemical experimentation, which he had practiced for most of his life and instead spent his time in meditation, in preaching, and in helping the poor.

After a time he left Switzerland to resume his wanderings. Refusing even to wear shoes, he lived by begging, and preached to the poor, while continuing to administer to the ill. Once, feeling completely exhausted, he sought refuge at Innsbruck, but the town had no use for "charlatans and vagabonds." Paracelsus wrote later, "The burgomaster of Innsbruck recognized doctors only in silk and at courts, not broiling in the sun in tattered clothes." So he continued his wanderings. Then, hearing that plague had broken out in the town of Sterzing, he hurried there to tend to the ill. But once the progress of the plague was halted, Paracelsus was ordered to leave the town. Later, in a letter to the town fathers, accusing them of ingratitude, he added a new title after his name: Professor of Theology.

Though Paracelsus scorned religious authority, once describing Martin Luther and the Pope as "two whores discussing chastity," he remained a Roman Catholic, if a somewhat unorthodox one. He supported the Church on the question of Christian unity and upheld much of Catholic dogma. He differed from the Church on several points, however. For example, he maintained that both heaven and hell were on Earth. According to Paracelsus, Adam and Eve were never driven out of the Garden of Eden. On the contrary, the place was changed from paradise to hell. Furthermore, he rejected the idea of life after death. There would be no resurrection. Humanity, however, had the ability to re-create paradise on Earth. In other words death was final. But the spiritual resurrection of humankind should be possible.

A TURNABOUT

After leaving Sterzing in the company of two friends in the town, Paracelsus's fortunes changed. The three went to Merano where Paracelsus "found much honor, happiness and fortune." He acquired patients in Merano, and soon he was no longer poor. But he was still not content to stay in any place for very long. He went to the spa of Pfeffer-Ragatz, near St. Moritz, where, at the request of a Benedictine abbot, he wrote a pamphlet about the virtues of the water from the springs. Then he revisited Schwatz, working for a while in the local mines.

In 1535 Paracelsus wrote the *Great Surgery Book*, though it wasn't really about surgery and didn't describe a single operation. Instead he discussed ways of avoiding them. In this book, Paracelsus criticized the practices of cauterizing wounds and of applying materials that infected rather than healed them. He emphasized that nature, not the doctor, was the healer and stated that, "Surgery consists in protecting nature from suffering and accident from without that she may proceed unchecked in her operations."

The book was published in 1536 with a preface by Wolfgang Thalhauser, the Augsburg municipal doctor. It was a huge success, and a second printing appeared in 1537. As his fame spread, Paracelsus, who had lived as a wandering beggar just a few years before, became wealthy. He was called in to treat members of royal houses and twice had audiences with King Ferdinand of Bohemia and Hungary, who had become interested in Paracelsus's ideas. But he spent the money that he received even faster than he earned it. He got into quarrels, sometimes over insignificant fees, and more than once became involved in litigation. One of his quarrels was with King Ferdinand, who had offered to pay the printing costs of one of his books. Paracelsus apparently squandered the money that was supposed to have gone to the printer, and the king branded him as a swindler.

In 1537 Paracelsus returned to Villach to take possession of the estate of his father, who had recently died. But the people of Villach were apparently not happy abut the idea of having Paracelsus as a resident, and the doctors of the town staged a demonstration against him during a church service that he was attending, demanding that he leave town at once. This time he did not engage in confrontation but went to St. Veith, some 20 miles away, where he had friends at the court of an archbishop and where he hoped to finally settle down. He was unable to get a permanent position at St. Veith. However, in 1540 the prince-bishop of Salzburg, Duke Ernst of Bavaria, offered him an appointment, and he spent his remaining days there. He died at Salzburg's White Horse Inn the following year after a short illness. The exact circumstances of his death are uncertain. According to one story, his enemies threw him off a cliff, or hired thugs to beat him as he walked down the street. According to another story, he fell downstairs while drunk. According to a third, the cause of his death was an overdose of laudanum. In any case, he knew he was dying and made a will three days before he died. He ordered that three psalms be sung for him in a church and that a coin be given to every poor man who came to the church door. He left 10 guilders to a relative, or ordered that 12 guilders be paid to his executors. He left his instruments and medications to the Salzburg barbers and listed places where his unpublished manuscripts could be found. Then on September 24, 1541, he died.

PARACELSUS THE ALCHEMIST

It would be wrong to view Paracelsus as the first chemist. He wasn't; he was an alchemist. Many of his ideas derived from alchemical tradi-

tions. For example, as I have noted previously, alchemists generally
believed that metals were composed of sulfur, mercury, and arsenic.
Gold was thought to be made up of sulfur and mercury alone.
Paracelsus held to a variant of this idea, maintaining that all matter,
including all metals that made up plants and the bodies of animals,
was composed of three "principles": mercury, sulfur, and salt. These
were not the familiar substances that we know, and they were more
like properties than elements. For example, salt was not the common
salt that we know, but rather a *principle* that conferred salinity.
Similarly, the mercury principle produced metallic luster, and the
sulfur principle the quality of combustibility.

According to Paracelsus, the traditional four elements were
receptacles for the principles. Like his contemporaries, he believed
that earth, air, fire, and water were the basic building blocks of the
universe. The principles, however, gave the elements the qualities they
had, or as Paracelsus put it:

> The world is as God created it. He founded this primordial
> body on the trinity of mercury, sulfur and salt and these are the
> three substances of which the complete body consists. For they form
> everything that lies in the four elements, they bear them all the
> forces and faculties of perishable things.

While Paracelsus often maintained that attempting to transmute
base metals into gold was a waste of time, he nevertheless attempted
the transmutation himself. He accepted the then-common idea that
metals "grew" inside the Earth and that, given enough time, all of
them would eventually become gold. If other metals transformed
themselves into gold naturally, then it seemed only reasonable that
the transformation could be made to happen by artificial means.

Another goal of the alchemists was to find the elixir of life, a
substance that could restore lost youth and prolong human life far
beyond the usual span, up to a thousand years, according to some.
Paracelsus never made such excessive claims, but he did administer a

preparation that he claimed could extend the span of life and rejuve-
nate the aged. This elixir, Paracelsus said, could be made by dissolving
pure potassium carbonate in water and then steeping some leaves of
the melissa plant in the solution. Alcohol was poured into the mixture several times. Finally the liquid was distilled and then boiled
down until it thickened into a syrup.

On the other hand, Paracelsus did make some legitimate chemical
discoveries, including a method for obtaining metallic arsenic from
arsenic sulfide, and he might have been the first to describe the
properties of two other metals, bismuth and cobalt.* Although he
was unable to break these substances down further, he failed to
recognize that they were elements. He couldn't have, not without
repudiating the alchemical theories that were common in his day.

Most importantly of all, Paracelsus was an experimental scientist.
He didn't content himself with just following traditional alchemical
recipes. Instead, he investigated the chemical makeup of the natural
world. He believed that God was a kind of Supreme Chemist. Thus,
by studying nature, one could gain knowledge of the workings of the
universe. "If Christ said, 'Investigate the Scriptures,'" he asked, "why
should I not say, 'Investigate the nature of things'?" Thus Paracelsus
became an experimenter at a time when, for all practical purposes,
experimental science did not yet exist. Yet he could accomplish only
so much. He was trying to understand the chemistry of nature at a
time when no one knew anything about the chemical elements. He
tried to find the laws of nature, but he had no terminology with which
to speak of them. He was able to describe the things he discovered
only in astrological or alchemical terms. No specifically scientific
language yet existed, and he was forced to use ill-designed concepts.

Paracelsus did not discover natural laws in the sense that Galileo
and Newton did. But like them, he sought to understand how nature
worked. He sought the key to the universe.

---

*There is some uncertainty about this. The discovery of cobalt is nor-
mally dated much later, in 1735.

CHAPTER 3

The Sceptical Chymist

The seventeenth-century English scientist Robert Boyle is generally considered to be the founder of modern chemistry. Before Boyle's time, chemistry was the province of the alchemists and of such craftsmen as dye makers and metallurgists. By the time Boyle died, chemistry had become a science.

Although it was Boyle's influence that caused the "al" to be dropped from alchemy, he was also an alchemist who spent years ardently seeking the Philosopher's Stone. He believed that he had a method for preparing philosophical mercury, which was thought by many to be a necessary step in creating the Stone. Boyle believed that he had witnessed a transmutation of lead into gold, and he claimed to have carried out a reverse transmutation of gold into silver himself. Though Boyle was skeptical of the long-held theory of the four elements, he could be quite gullible when hearing tales of alchemical transmutations or of magic. A founding member of England's Royal Society, Boyle also allowed himself to be inducted, at one point, into what was supposedly a society of alchemical adepts.

Robert Boyle was born in 1627, the youngest son of Richard
Boyle, the first earl of Cork, who was then the wealthiest man in the
British Isles. The elder Boyle had gone to Ireland at the age of 22 to
make his fortune. And he did exactly that, acquiring vast estates in
Ireland and setting up ironworks and other industries. Although he
amassed enormous wealth, in some respects Boyle had a checkered
career. In 1598 he was imprisoned on charges of embezzlement. However, he was allowed to present his case in person to Elizabeth I and
her Privy Council and was acquitted. In 1620 he was appointed a lord
high justice, and he became lord high treasurer in 1631. But in 1633
he had to face charges that he possessed defective titles to some Irish
estates he had purchased from Sir Walter Raleigh and was forced to
use bribery—including a "loan" of £14,000 to Charles I—in order to
remain in possession of his lands.

The earl of Cork believed that children were best brought up away
from home until it was time to begin their education. So, shortly after
his birth, Robert was put in the care of a peasant woman until he was
old enough to be taken back home and instructed by tutors. At the
age of eight he was sent to Eton, where he remained for three years.
He then returned home and was again taught by tutors.

GRAND TOUR

In 1639 Robert and his older brother, Francis, were sent to Europe in
the company of a tutor. The three went first to France and passed
some time in Paris before going on to Geneva, where their education
was to continue. All in all, they spent some 21 months in Geneva. The
two boys studied Latin and rhetoric and read Roman history in
French. Part of each day was set aside for reading and discussing the
Old and New Testaments, and the afternoons were devoted to such
activities as fencing and tennis.

By 1641 the two boys had grown tired of life in Geneva and asked
their father for permission to travel in Italy. Permission was granted,

and in September of that year they crossed the "hideous mountains," the Alps, that separate Switzerland from Italy. They passed the winter in Florence. In 1642, while Boyle was in Florence, Galileo died in Arcetri, a nearby town. Boyle began to read Galileo's scientific works and soon became a convert to his scientific philosophy.

In the spring the boys went to Rome and then to Genoa by way of Florence. From Genoa they sailed north to France, and then went on to Marseilles, where they expected to get money, sent by their father, that would allow them to return to England. They waited for days, but no money came. Eventually, in May of 1642, they received a letter from their father. According to the earl, the Irish had rebelled against the English, and he was defending himself in fortified castles. The costs of the war had depleted his ready funds. However, he had obtained £250 by selling some plate and had sent bills of exchange for that amount to his agent in London.

But no bills of exchange were forthcoming from the agent. The boys and their tutor debated what to do. They still had reserve funds, enough to pay for some further travel. Francis insisted on returning to Ireland to fight alongside his father and brothers. But Robert had no desire to be a soldier; he decided to go back to Geneva with his tutor. Boyle lived there until 1644, supported by the tutor. Finally, after exhausting what little money remained, he sold "some slight jewels at a reasonable rate" and returned to England.

RELIGIOUS VOWS

In 1640, when he was 13 years old, Boyle experienced a religious crisis that was to profoundly affect the course of his life. One summer night, while he was in Geneva, he was awakened by a violent thunderstorm. The flashes of lightning were so frequent and the thunder so loud that he imagined that the day of judgment was at hand. Feeling unprepared to face God, he vowed that, if he lived through the night, ". . . all his future additions to his life should be more religiously and watchfully employed."* When morning came and revealed a cloud-less sky, he reaffirmed his vow.

Other adolescents might have quickly forgotten such a promise, but not Boyle. From that day on, he did his best to live a life of Christian piety. He never used expletives and always paused reverently before saying the name of God. He also practiced lifelong celibacy. When, as an adolescent, he was taken by his tutor to the brothels of Florence, Boyle resisted any desires he might have had to sample the pleasures offered there. Years later he wrote that he considered the sights that he encountered in the brothels to be convincing arguments for chastity.

When he was a little older, Boyle began to make himself into a Biblical scholar. In addition to Latin, which all educated people of his day knew, he learned Greek, Hebrew, Aramaic, and Syriac (a dialect of Aramaic) so that he could read sacred texts in their original languages. He wrote extensively on religious subjects throughout his life, and in his will he endowed the Boyle Lectures, a series of eight annual sermons "for proving the Christian Religion against notorious Infidels." Boyle was convinced that atheism was a threat to Christianity and believed that the lectures would combat atheism. The Boyle Lectures are still given today.

Believing that Roman Catholicism and Islam were heresies almost as bad as atheism, Boyle had the Bible translated into Gaelic and Turkish. The Gaelic Bible was distributed throughout Ireland and Wales and also in the Scottish highlands. When Parliament passed an act establishing a society for the propagation of the Gospel among the Native Americans of New England in 1649, Boyle accepted the position of governor of the society. He took the post quite seriously and he left large sums to Harvard College and William and Mary College in his will to spread the Christian faith in the New World.

During Boyle's lifetime, one of his better-known works was *Seraphic Love*. This was written in the form of a book-length letter to

---

*Here Boyle is writing about himself in the third person.

a fictional friend who had been jilted by a woman and in it Boyle briefly expounds on the uncertainties of erotic love, then urges his friend to devote himself to a higher kind of love, the seraphic love of God. He wrote the book in his youth and originally did not intend to publish it. It was meant only for private circulation. When one of the privately circulated copies was offered to a printer, Boyle decided to publish the book himself. Though *Seraphic Love* is rarely read today, it was a great success in its own day. Between 1659 and 1708 there were nine English editions, and the book was translated into French, German, and Latin.

Another religious romance by Boyle, *Martyrdom of Theodora*, was less successful. Nevertheless, it seems to have given birth to a trend, because the writing of religious novels soon became something of a fad in England. Indeed, the English literary scholar and critic Samuel Johnson wrote that, "The attempt to employ the ornaments of romance in the decoration of religion was, I think, first made by Mr. Boyle's *Martyrdom of Theodora*; but Boyle's philosophical studies did not allow him time for the cultivation of style."

Boyle did not confine himself to writing religious fiction. Throughout his life he published books and essays on religious topics. Among them were works with such titles as *The Christian Virtuoso, Excellency of Theology, Free Discourse Against Customary Swearing, Protestant and Papist, Some Considerations Touching the Style of the Holy Scriptures,* and *Some Physico-Theological Considerations About the Possibility of the Resurrection.* Boyle was not only a natural philosopher, alchemist, and novelist, but also a theologian, whose theological writings were held in high regard. Thus he was offered high positions in the Anglican Church more than once. Characteristically, he refused them, saying that he thought he could be more influential as a layman.

LONDON

When the 17-year-old Boyle arrived in London in 1644, his prospects did not seem to be very good. He had little or no money and no friends in the English capital. Civil war had broken out in England between Charles I and the parliamentarians. Peace had been negotiated in Ireland, but the earl had died the previous year, and Boyle's family knew nothing of his whereabouts.

According to Boyle, shortly after he arrived in London, he "accidentally" met his sister Katherine, Viscountess Ranelagh. Boyle quite naturally attributed this fortunate occurrence to the hand of providence. But perhaps one should not take this account too seriously. There is reason to think that Boyle's tutor provided him with a letter of introduction to Dr. Theodore Diodati, a well-known London physician. If he did, then Dr. Diodati would certainly have arranged a meeting between Boyle and his sister.

In any case, Boyle lived at his sister's house for the next four and a half months. During that time he was introduced to a lot of people, including a sister-in-law of Lady Ranelagh who happened to be married to a prominent member of the so-called Long Parliament This influential M.P. used his connections to secure Boyle's estates in England and Ireland and arranged a safe conduct that allowed Boyle to travel to one of them, at Stalbridge in the county of Dorset.

During his years at Stalbridge, Boyle devoted a great deal of his time to religious writings and in nursing his real and imaginary ills. It was there that he wrote *Seraphic Love* and *Martyrdom of Theodora and Didymus*, for example. Meanwhile, he suffered from frequent indigestion, malarial fevers, colds, and poor eyesight. From the age of 20, he was also afflicted with kidney stones. Boyle coddled himself in numerous different ways. He limited himself to a simple diet and, according to Thomas Birch, a contemporary who wrote a biography of Boyle, "he had diverse sorts of cloaks to put on when he went abroad, according to the temperature of the air; and in this he governed himself by his thermometer."

Boyle was a firm believer in iatrochemistry, the preparation of drugs by chemical means ("iatro" means medicine or healing). He devoted a great deal of time to trying to develop new medicines and to collecting medicinal recipes from every source he could find. He then tried out these medicines on himself, rating them "A," "B," or "C" according to how effective he thought they were. This could not have been good for his health, because some of the preparations contained compounds of mercury and antimony, which are poisonous.

Fortunately, some of the remedies that Boyle used or recommended did no harm, because they were not taken internally. For example, when he experienced an attack of malaria, Boyle placed a small packet containing a local plant called groundsel on his stomach. He believed that jaundice could be treated by placing a sheep's gallbladder over the patient's bed. As the gallbladder dried up, the jaundice would disappear "in sympathy." And he had a ring carved from an elk's hoof that he kept by his bedside as a remedy for the stomach cramps that he often experienced. It is likely that such treatments did make Boyle feel better through a placebo effect. And they probably had the added benefit of causing him to ingest fewer poisons.

OXFORD

While living at Stalbridge, Boyle became increasingly interested in chemistry. He set up a laboratory in his house and quickly became absorbed in chemical and alchemical experiments. Although this work produced no significant discoveries, it convinced him of the importance of studying nature empirically. A somewhat religious fervor characterized his experimentation. He regarded science, not as an end in itself, but as a means of discovering the nature and purpose of God.

In late 1655 or early 1656, Boyle moved to Oxford, where he could enjoy the company of other natural philosophers. In 1645 a number of people interested in the new experimental philosophy had formed a group, called the Invisible College, which met weekly in London. However, the Civil War had interrupted these meetings. Some of the members had migrated to Oxford, which was a royalist stronghold at the time; others had left the capital city for a variety of reasons. By 1655 some of the members of the Invisible College had become affiliated with Oxford University, and the weekly meetings resumed. On one of his visits to London, Boyle met Dr. John Wilkins, under whose leadership the Invisible College had originally been created. Wilkins, who had become Warden of Wadham College, urged Boyle to move to Oxford, where he could participate in the group's activities.

During his years in Oxford, Boyle engaged in an intense program of experimentation and writing. Beginning in 1660, he wrote a series of books on different aspects of natural philosophy and contributed to the *Philosophical Transactions* of the Royal Society, which was established in 1660 as a successor to the Invisible College. It was during his Oxford period that Boyle did the work that caused him to become the best-known natural philosopher in England. He would be overshadowed only by Isaac Newton.

Boyle brought a new kind of rigor to writing about natural philosophy. Experimentation was nothing new; most natural philosophers engaged in it. But they commonly omitted the details that would make it possible to duplicate their experiments. Furthermore, they frequently fudged their results, making them seem better than they actually were. This wasn't considered fraudulent at the time. Even Galileo engaged in this practice and sometimes went so far as to describe experiments he had never performed. Apparently Galileo believed that, because he already knew how certain experiments would turn out, it wasn't necessary to actually do them.

Boyle, on the other hand, described his experiments in great detail so that other scientists could repeat them and, in some cases, extend them. Boyle's example helped to change the ways that experiments were reported, not only in England but throughout Europe. He had much of his work translated into Latin so that non-English scholars could read it. So influential was Boyle, if we are to believe one of his publishers, that he became known in continental Europe as "the English philosopher." Nevertheless, he often neglected to describe his experiments in the order in which they were performed. His writing was often muddled, and he had a habit of correcting and adding to manuscripts after they were already in the hands of his printers. Many of his books are basically assemblages of bits and pieces, perhaps partly a result of his poor eyesight. By this time Boyle could no longer see well enough to write and had to dictate instead. Furthermore, he was often unable to read what he had dictated. That couldn't have been conducive to the production of clear and concise writing.

THE SPRING OF THE AIR

When Boyle began his scientific work, chemistry was not yet a science. The complexity of the materials with which chemists worked made generalizations about their behavior difficult. There was no standard chemical terminology and no concept of chemical purity. Paracelsus had tried to use chemical preparations that were as pure as possible but most who succeeded him didn't bother. No distinction was made between organic and inorganic substances, and there was no clear understanding of the difference between acids, bases, and salts.

One of the greatest impediments to progress was the lack of knowledge about the composition and nature of gases. Chemists confined themselves to working with solids and liquids and paid no attention to the gaseous state of matter. This sometimes led to wildly erroneous conclusions. For example, the seventeenth-century Flemish nobleman Joan-Baptista van Helmont performed an experiment in which he placed the 5-pound stem of a willow tree in a vessel containing 200 pounds of earth that had been dried in an oven. Over the course of five years, the stem, watered only with distilled water and rainwater, grew into a tree weighing 169 pounds. Van Helmont dried the earth in which the tree had been growing and weighed it again. He found that the earth now weighed only two ounces less and concluded, "Hence one hundred and sixty-four pounds of wood, bark and roots had come up from water alone." It never occurred to him to consider the fact something in the air might have had anything to do with the growth of the tree. Indeed, he had no way of knowing that plants take up carbon dioxide; no one knew what carbon dioxide was in his day. He was blind to the role that gases could play in chemical reactions. But then so were most of the other natural philosophers of the day. It took a spectacular public demonstration to awaken their interest in air.

Scientific experiments are almost always performed in secluded laboratories. However, a few were intended as public spectacles. One such experiment was performed by the German engineer Otto von Guericke before the Holy Roman Emperor Ferdinand III and a large crowd that had come to watch. Guericke, who had invented an air pump, employed a blacksmith to pump the air out of a sphere consisting of two large copper hemispheres that had been fitted together. Two teams of eight horses were then brought forward and one team was harnessed to each of the two hemispheres. The 16 horses proved unable to pull the hemispheres apart. Yet the hemispheres fell apart spontaneously as soon as Guericke let the air back into them. He explained that it was the pressure of the surrounding air that had held the hemispheres together.

News of Guericke's experiment quickly spread throughout Europe. Boyle was very impressed when he heard of it. He realized that, with a good pump, he could perform experiments to determine the properties of air. He was fortunate to have an assistant, Robert Hooke, who was a talented experimenter and designer of experimental equipment. With the aid of Hooke, he built a vacuum pump that was an improvement on the device that Guericke had used, and he and his assistants performed numerous experiments using it. In one a watch was suspended in a container, and as the air was pumped out of the container, the sound of the ticking grew fainted and fainter, demonstrating that sound could not pass through a vacuum. In another a feather was observed to fall straight down in a vacuum with none of the fluttering seen when a feather fell in air. In yet another a lamb's bladder was placed in a container. The bladder's neck was tied to prevent what little air it contained from escaping. As the air was pumped out of the container, the air inside the bladder expanded. When the air was let back in, the bladder shrank to its original size. One of the most important experiments was one in which a mercury barometer was placed with its base in a glass globe. When the air was pumped out of the globe, the mercury column of the barometer fell, and then rose again to its original height when the air was let back in. This confirmed the hypothesis that the mercury column was sustained by air pressure and demonstrated that air had weight.

These and other experiments were fully described in Boyle's book *New Experiments Phisico-Mechanicall, Touching the Spring of the Air, and Its Effects (Made for the Most Part in a New Pneumatical Engine)*, which was published in 1660 and established his reputation as a natural philosopher. *Spring of the Air* had a great influence on other scientists, who built their own vacuum pumps, repeated Boyle's experiments, and devised new ones of their own. Air, they now realized, was something that had very specific properties that could be experimentally investigated.

Boyle's experiments with the vacuum pump ceased in 1662, when Hooke moved to London to become curator of the Royal Society. However, by this time, Boyle and Hooke had performed a number of additional experiments that are described in the second edition of *Spring of the Air*, published in 1669. Included in the second edition was a statement of what is now known as Boyle's law: the pressure of a gas is inversely proportional to its volume. In other words, if the pressure is doubled, the gas is compressed to one-half of its former volume and if the pressure is halved, the volume doubles.

Boyle's law isn't exact. It ceases to be valid at very high pressures and at low temperatures, and it remains correct only if the temperature of the gas is held constant. However, the experiments that led Boyle to the law showed that the properties of air could be described in a mathematical way. This in itself was an important step forward. It also led Boyle to conclude that air must be made up of tiny corpuscles. If this was the case, then increasing pressure could force them closer together, while decreasing pressure allowed them to move farther apart. And if air was composed of corpuscles, it seemed likely that solids and liquids were, too.

THE SCEPTICAL CHYMIST

The Sceptical Chymist, which was published in 1661, is the most famous of Boyle's works and a science classic. It is probably also one of the least read. Unlike Boyle's descriptions of his chemical experiments, which were generally very clear, *The Sceptical Chymist* is practically unreadable, so much so that some historians of science prefer to quote from an early draft of the book rather than from the published text. *The Sceptical Chymist* has numerous repetitions, and the writing is often digressive and muddled, and sometimes self-contradictory. It has all the earmarks of a work that was hastily thrown together. Boyle himself admitted that the book was "maim'd and imperfect." Curiously, the original 1661 edition has two title pages, one at the front of the book and one following page 34. They both have the same title, but different subtitles that describe the book differently.

Although Boyle uses the word "chymist" rather than "alchymist," the book is no condemnation of alchemy. By the time he wrote the book, Boyle was a practicing alchemist himself, and he did not make the modern distinction between chemistry and alchemy. In fact, in the second edition of the book, which was published in 1680, Boyle speaks quite favorably of the alchemical adepts.

The Sceptical Chymist, which is written in the form of a dialogue between five people (two of whom mysteriously disappear in the part following the second title page and reappear near the end of the book), is a discussion of the chemical philosophies that prevailed in Boyle's day, the Aristotelian theory of the four elements and Paracelsus's theory of three principles. Boyle discusses them in exhaustive detail in order to foster skepticism concerning them.

At the time, it was commonly believed that *all* bodies contained all four of the elements and furthermore that substances could be broken down into their components by fire. However, Boyle said, some bodies produced fewer than four substances when they were heated, and some produced more. Observation indicated that the idea that there were four basic substances was false. Furthermore, some substances, such as gold, had never been analyzed into components.

A considerable part of the book is an attack on the Paracelsian theory of three principles. Boyle argues, for example, that none of the principles can be used to explain how a prism breaks up white light into different colors. Furthermore, authors who argued in favor of the three principles generally described them in such obscure ways that it was doubtful that they understood these supposed principles themselves.

Boyle expounded no alternative theory in *The Sceptical Chymist*. He concerned himself mainly with fostering skepticism about the Aristotelian and Paracelsian chemical philosophies. But the book is a science classic nevertheless. Little progress could be made in chemistry until those theories were overthrown. To be sure, Boyle failed to accomplish this and belief in the four-element theory, especially, lingered on for quite a long time. However, Boyle showed that it was possible to doubt long-established ideas, thus performing a great service to science.

A "LABORIOUSLY USELESSE" CONCEPT?

Like many natural philosophers of his day, Boyle was an atomist. He summed up his atomistic hypothesis in *Origine of Forms and Qualities*, published in 1666, in which he stated his belief that there was one kind of "Catholick or Universal Matter" which existed in the form of tiny corpuscles of different sizes, shapes, and motions. These properties of the corpuscles caused different chemical substances to have different properties. For example, noting that nitrate crystals were prismatic, Boyle reasoned that the corpuscles that made up the crystals were tiny prisms. He speculated further that it is the sharp ends of these crystals that cause nitric acid to be corrosive. Of course he was totally wrong, but his arguments showed that the causes of the particular properties of different substances were open to speculation. Boyle also suggested that different kinds of corpuscles could bind together in small clusters. This time he was absolutely correct. There is no significant difference between Boyle's idea and the modern conception of a molecule. Furthermore, the idea that different kinds of corpuscles, or atoms, could combine with one another was the first step toward understanding the nature of chemical reactions.

When Boyle published the second edition of *The Sceptical Chymist* in 1680, he added an appendix that gave his definition of a chemical element. He wrote:

> I now mean by elements, as those chymists that speak plainest do by their principles, certain primitive and simple, or perfectly unmingled bodies; which not being made of other bodies, or of one another, are the ingredients of which all those perfectly mixt bodies are immediately compounded, and into which they are ultimately resolved.

This was not first time that the idea of a chemical element was defined. Nor does it seem, at first glance, to be anything that would really advance the science of chemistry. All Boyle was saying, after all, was that an element was anything that was not a compound or a mixture of different substances. What gave the idea importance was Boyle's insistence that only experiment could determine what was or was not an element. If a substance could be broken down, then it clearly was not an element. One should not decide, for theoretical or philosophical reasons, that certain substances (for example, earth, air, fire, and water) were the elements of which everything else was composed.

Boyle himself did not consider his definition of a chemical element to be especially important. In fact he called the concept "laboriously uselesse." Perhaps it was not a very useful concept in Boyle's day. There was really no way to determine what the true chemical elements were. However, the idea would eventually prove to be of great importance for the development of chemistry.

LONDON

In 1668 Boyle moved back to London, to live with his sister Lady Ranelagh, but there was no interruption in his chemical experimentation. Even after suffering a paralyzing stroke in 1670, he continued to work, directing the work of his assistants by dictating instructions from his bed. Somewhat surprisingly, some of his best experimental work was done after the stroke.

I haven't devoted much space to discussions of Boyle's experimental work, because his theoretical ideas influenced chemistry much more, so perhaps it won't hurt to mention a few of his experiments here. He investigated the nature of acids and alkalis and discovered an indicator that allowed one to tell whether a solution was acidic or alkaline. This indicator was a liquid known as syrup of violets, which was prepared by boiling violet petals in water and adding sugar. Boyle found that acids turned the syrup red and alkalis turned it green.

He developed other tests that could be used to detect the presence of certain elements, such as copper or iron (Boyle didn't know that copper and iron were elements, of course). After being shown samples of the newly discovered element, phosphorus, Boyle and his assistants discovered how to obtain it from urine, and he described the properties of phosphorus so extensively that two centuries passed before there was anything to add.

Boyle and his assistants studied the nature of respiration, discovering, for example, that animals died in a vacuum but that insects did not. He concluded that something in the air was absorbed during respiration and during combustion as well. Naturally he was unable to identify this something as oxygen, because oxygen had not yet been discovered. Finally, he made a mass of observations on chemical substances and chemical reactions, publishing detailed accounts that any other chemist could follow. These accounts, because they were models of precision and detail, contributed greatly to the advance of chemistry.

Though Boyle had conceived the idea of a molecule, he invented no theory that explained chemical reactions. However, it was not possible to go any further than Boyle did because much more had to be discovered before a real understanding of the nature of chemical processes could be obtained.

BOYLE THE ALCHEMIST

Boyle first became interested in natural philosophy during the late 1640s, and he was intrigued by the idea of alchemical transmutations from the very first. From the time he set up a laboratory in Stalbridge to the end of his life, the pursuit of the Philosopher's Stone was always one of his preoccupations. Boyle's personal library contained numerous books on alchemy, and a great many manuscripts describing alchemical experiments were found among his papers after his death.

Like many other alchemists, Boyle distinguished between two kinds of transmutation, particular transmutation and projection. "Transmutation" was any conversion of one metal into another. It was held to be far easier than "projection," which meant using the Philosopher's Stone to transmute a base metal into gold. It was thought that the Stone could transmute a quantity of metal that was hundreds or thousands of times greater than the Stone's weight.

Boyle believed that he had carried out a transmutation himself. In *Origine of Forms and Qualities*, he describes a reverse transmutation of gold into silver. He prepared a corrosive solvent by mixing nitric acid and butter of antimony (antimony trichloride). According to Boyle, when this preparation was poured over gold, the gold dissolved, producing a white powder, which when fused with borax produced small metallic globules. Boyle took these globules to be silver.

Boyle described this transmutation clearly and in detail. But the manuscripts that describe his attempts to find the Philosopher's Stone and his letters to other alchemists are written quite differently. They are always in code. He used a variety of ciphers and invented code words for different chemical substances or alchemical procedures. Like most other alchemists, he believed that if methods of preparing the Philosopher's Stone became widely known, the results would be catastrophic.

Boyle did not publish any works about traditional alchemy, but a number of unpublished works on alchemy were found among his papers after his death. One of these, *Dialogue on Transmutation,* exists today only in fragmentary form. It is especially interesting because it contains a firsthand account of a transmutation by projection. Boyle apparently believed that he had seen a "forraigne doctor of Physick" (who is not named) change lead into gold by the Philosopher's Stone. The foreigner is said to have thrown a small quantity of a red powder into some molten lead. He covered the crucible in which the lead had been melted and the crucible was heated for 15 minutes and then cooled. When the contents of the crucible were removed, they were found to be gold.

Dialogue on Transmutation is just that, a dialogue. In it Boyle never says that *he* witnessed the foreigner's projection. However, the same incident is described in notes taken by Boyle's confidant, Bishop Gilbert Burnet, when Boyle told him of the incident. And Boyle refers to the incident himself in his book *Producibleness of Chymical Principles.* Here he referred to the operation carried out by the foreigner only as a "Metalline experiment," not as a projection. However, the details that he recounted were the same as those in the *Dialogue.*

Boyle might have witnessed other projections as well but whether he did or not, he was clearly convinced that the transmutation of lead into gold was possible. This belief was not inconsistent with his chemical philosophy. Recall that he believed that the atoms, or corpuscles, of which all substances were composed were made of the same kind of primal matter. It followed that if there were ways to change the sizes and shapes of these corpuscles, transmutations could be carried out.

PHILOSOPHICAL MERCURY

Boyle had numerous contacts with other alchemists, among them the American George Starkey. Starkey, who was born in Bermuda, performed alchemical experiments for years after graduating from Harvard in 1646 and moved to England in 1650. When he arrived, Starkey had several manuscripts supposedly written by a New England adept who used the pseudonym Eiraneus Philalethes. Starkey claimed to have witnessed several transmutations that Philalethes carried out and had seen him use the Stone to restore a withered peach tree and to make an old woman grow new teeth.

Of course Philalethes never existed. He was entirely fictional, a product of Starkey's imagination. However, during the seventeenth century there was no lack of people willing to accept miraculous stories on faith, and Boyle was one of them. Shortly after Starkey was introduced to Boyle in the early 1650s, the two began to collaborate on chemical and alchemical projects. In 1651 Boyle received from Starkey a letter describing the preparation of philosophical mercury.

You may recall that the preparation of philosophical mercury was one of the steps toward creating the Philosopher's Stone. Boyle must have believed that he had learned a great secret indeed. He must have become even more excited when he discovered that heat was produced when this "mercury" was used to dissolve gold. According to alchemical theory, the heat was a sign that some of the mercury was being transmuted into gold.

Boyle described this substance in a 1675-1676 issue of the *Philosophical Transactions* of the Royal Society. In this paper, which was published under the reversed initials "B. R.," Boyle does not reveal how his "incalescent mercury" was made and does not identify it as philosophical mercury. He says only that philosophical mercury is "of kin" to the substance he prepared. But he describes the production of heat and explains how other "mercuries" can be tested to see if they produce heat with gold, too.

Boyle's use of the plural term "mercuries" might seem strange to the modern reader, but it would not have been thought odd in his day. The alchemists used the term "mercury" to describe many different substances, and many thought that a different mercury was associated with each metal. There is some uncertainty about exactly what Boyle's incalescent mercury was. It was probably prepared from the metal, however. Alchemists generally considered philosophical mercury to be a highly purified form of the common metal.

Boyle was determined to keep his method of preparation secret. He published the paper on condition that no questions be asked about its contents. He later reiterated his determination to maintain secrecy, saying that he would never make this mercury again or reveal how it was prepared. This was a vow that Boyle didn't keep, incidentally. He later shared Starkey's recipe with various alchemical collaborators.

BOYLE THE ADEPT

During the summer of 1677, Boyle received a letter from a man named Georges Pierre des Clozets. Georges Pierre sent a copy of a letter signed by "Georges, patriarch of Antioch" that ordered him to go to England and visit Boyle and demonstrate "the projection." Georges Pierre was also to tell Boyle that "it will not be long before God allows him [Boyle] the happiness of being a true philosopher." Boyle must have been impressed by his French visitor, because he immediately began a long correspondence with him. No letters from Boyle to Georges Pierre exist today. However, some 13 of Georges Pierre's letters to Boyle survive and they refer to letters that Boyle had written him.

Then in early 1678 Boyle got a letter from a man who identified himself as Georges du Mesmillet, the patriarch of Antioch, and the head of a society of alchemical adepts. In this letter the "patriarch" acknowledges Georges Pierre as his agent and tells Boyle that he will be nominated for membership in a secret society of alchemical masters. Boyle is informed that members of the secret society know of Boyle's accomplishments as a natural philosopher from his books and that they have heard reports of his piety and generosity.

Boyle continued to write numerous letters to Pierre and also sent Pierre gifts for himself and for the patriarch. Pierre then asked Boyle for more gifts, ones that the patriarch could present to the Turkish court. Boyle was asked to send, among other things, jackets, a chiming clock, and "for the sultana queen mother, eight rods of flesh-colored moiré, eight of gold-colored moiré, and eight of flame-colored moiré." Both Georges Pierre and the patriarch promised Boyle gifts in return.

Because Boyle's letters are lost, there is no way of knowing whether Boyle received any gifts other than some fruits and cheeses sent by Georges Pierre. But the list of items he was told he would receive is certainly impressive. It included "two excellent pearls, gold brocade, satin, Chinese porcelain, silk carpets, and gold ingots." Boyle was also advised to become a member of the East India Company to facilitate his receipt of gifts from the East. When Boyle replied that he was already a member, Georges Pierre told him "to continue what you have begun, to do what is necessary for you to belong to the Company of Turkey." One of the gifts promised Boyle was a powder of projection that could be used to turn base metals into gold, in other words a sample of the Philosopher's Stone. We can be certain that this never arrived because in one of his letters the patriarch expresses regret that it was lost in transit.

In March of 1678 Georges Pierre wrote Boyle that the members of the secret society had assembled near Nice. Then, a week later, he informed Boyle of some of the events that had taken place at this meeting. One adept had demonstrated a powder that could transform water into a transparent stone. A "Polish philosopher" had caused plants to flower and to bear fruit in two hours. A "Chinese gentleman" showed some flasks that contained a homunculus, a five-month-old foal and a fox. Meanwhile, the assembly was expecting the arrival of three adepts "from the banks of the Ganges."

Shortly afterward, Boyle received a document that had been drawn up "at Herigo, in the county of Nice on the 24th of February." The document states that Boyle's nomination was accepted, along with those of two other applicants. The second half of the document, dated March 7, 1678, nominated Boyle to the post of treasurer for France, England, and Spain and reminds him that he must present himself before the society on May 6. In the event of illness, however, Boyle might send any of three men (one of whom was Georges Pierre) as his representative.

It is interesting to speculate what might have happened if Boyle had tried to attend the assembly in person. No such place as the castle of Herigo (the adepts were supposedly meeting in a castle) has ever been identified. But Boyle chose to send a representative, sending Georges Pierre 600 livres that he had requested to "defray expenses." Boyle was informed that his representative would "carry out all necessary functions" and that he could present himself before the society at a planned fall meeting, which would take place in London.

In June, Georges Pierre wrote to Boyle that a great chest holding a book in which was written "the true interpretation of all our emblems and all our formulations, which have been employed by the inhabitants of the chemical mountain to hide their foliated earth from unbelievers, the sworn enemies of God, and the allegories, parables, problems, types, enigmas, sayings of nature, fables, portraits and figures of the foster children of Nature" would be delivered to Boyle in London. Georges Pierre also asked Boyle for another 800 livres. It was one of the last letters that Boyle received because shortly thereafter, Georges Pierre disappeared. Naturally Boyle made inquiries about him, but was unable to get any definitive information.

Then, in late 1679 or early 1680, Georges Pierre returned to the French town of Caen, where he had previously resided. He was riding "a horse worth sixty livres" and had a servant equally well mounted. He had "the richest clothing and other furniture and rare and precious curiosities," according to a letter written to an unknown addressee in London. Georges Pierre bought an estate worth 14,000 livres and lived "in great splendor and opulence." That spring he began building and planting on his estate, but soon fell ill and died.

COMMUNICATING WITH SPIRITS

Boyle was above all a religious man. Furthermore he was wealthy. Because he believed that atheism was a threat to Christianity, he longed to discover ways that atheistic ideas could be refuted. In particular, he thought that if it were possible to communicate with spirits, the existence of a spiritual world, and hence of God, would be demonstrated.

Boyle believed that spirits or angels could be invoked with the Philosopher's Stone. This was not as eccentric a belief as it seems; it was an idea widely accepted by alchemists of his day. He believed, furthermore, that the spirits attracted by the Stone might be the cause of transmutation. Projection might be an interaction of the spirit world with matter.

Naturally, Boyle's hope of demonstrating a connection between the mundane world and the spirit world never became a reality. When he died he was no nearer finding the Philosopher's Stone than any of his predecessors had been. But he probably never stopped trying. After Boyle's death, the philosopher John Locke, one of the three people appointed by Boyle to sort through his papers and manuscripts, found some of his incalescent mercury and a "red earth" among his effects. Locke was deeply interested in alchemy and he had often shared alchemical secrets with Boyle and with Isaac Newton so he willingly sent samples of these substances to Newton at his request. Newton, who also practiced alchemy, would certainly have been aware that the Philosopher's Stone was often described as a red powder. Newton asked Locke for some coded papers describing an alchemical procedure involving Boyle's mercury. After he received them, Newton wrote, "This receipt [recipe] I take to be that thing for the sake of which Mr. B. procured the repeal of the Act of Parliament against

Multipliers." Here Newton was referring to the legislation enacted during the reign of Henry IV that outlawed the transmutation of metals into gold and silver. Boyle had been instrumental in getting it repealed.

What was the mysterious "red earth"? No one knows. We only know that, whatever it was, Boyle did not use it to make gold. Nor did Newton. Boyle died on December 30, 1691, exactly one week after the death of his sister. He was buried in the church of St. Martin's-in-the-Fields, which was later demolished, and no trace of his burial place exists today.

CHAPTER 4

THE DISCOVERY OF THE ELEMENTS

A t the beginning of the seventeenth century, 13 elements were known. Nine—carbon, sulfur, iron, copper, silver, gold, tin, lead, and mercury—had been discovered in ancient times. Four more—arsenic,* antimony, bismuth, and zinc—were discovered between around 1250 and 1500. It is not by chance that 11 of the 13 are metals. Some of them have relatively low melting points and were undoubtedly first produced when fires were laid on surface ores. Fires built by preliterate peoples in modern times have often produced small quantities of metals. A rich vein of silver was discovered in this manner by an Indian sheepherder in seventeenth-century Peru who built a fire at nightfall and found the next morning that the stone under the ashes was covered with silver.

Other metals, such as iron, have relatively high melting points. But iron can be smelted in fairly primitive furnaces, and it was known

---

*The Greeks and Romans knew a substance they called "arsenic," but this was an arsenic compound, not the metal.

in Neolithic times. Iron was known long before bronze—an alloy of copper and tin. But it initially wasn't used for weapons. Unalloyed iron isn't as strong as bronze, and it won't hold a sharp edge. In Homer's *Iliad,* for example, the heroes have bronze armor and use bronze swords. Nevertheless, iron was considered valuable. Achilles awards a lump of iron as a prize to the individual who can throw it the farthest.

The two non-metals, carbon and sulfur, have probably been known as long as human beings have known how to make fire. Carbon in the form of charcoal is a byproduct of fire and was used to make drawings on the walls of caves. Sulfur is found near volcanoes in the form of brimstone. It, too, was used in very early times. For example, after slaughtering Penelope's suitors, Odysseus fumigates his house by burning sulfur.

None of the 14 was known to be an element in 1650, the mid-point of the seventeenth century. All were supposedly mixtures of earth, air, fire, and water. One, tin, was generally supposed to be a mixture of silver and lead. The gases that were most abundant in the atmosphere, nitrogen and oxygen, remained unknown, as did the elements that are most abundant in the Earth's crust: oxygen, silicon, and aluminum. Chemists were ignorant of the composition of most of the substances they used. For example, nitric acid (then called *aqua fortis*) can be made by dissolving a compound of nitrogen and oxygen in water. But the components of this compound could hardly have been identified by individuals who didn't know what hydrogen, nitrogen, and oxygen were. Sal ammoniac (ammonium chloride) was a compound containing nitrogen, hydrogen, and chlorine. Again, all three of these gases were unknown. Furthermore, there was often confusion about the known chemical substances. For example, to the alchemists "sulfur" could be anything combustible.

Clearly, little progress could be made in chemistry until chemists gained a better understanding of the materials they worked with. Unfortunately, little progress was being made. In 1650 no new elements had been discovered for 150 years. The concept of a chemical element, in the modern sense of the term, was unknown. And when Robert Boyle defined "chemical element" in 1661, he pronounced the idea to be "laboriously uselesse."

GOLD FROM URINE

The next step forward was made not by a chemist but by an alchemist who was trying to make gold. A new element, phosphorus, was discovered by a Hamburg alchemist named Hennig Brandt (or Brand) who was trying to extract gold from human urine. Brandt's date of birth is unknown, and little is known about his early life. The best guess seems to be that he was born during the 1620s. We know that Brandt served as a soldier during the Thirty Years War, which raged across Europe during the years 1618 to 1648. After returning to civilian life he set himself up as a physician and insisted on being called "Herr Doktor," although he had no medical degree. It is probable that Brandt married around this time, but there are no records of his marriage. We know only that the Brandts had at least two children before his wife died.

Brandt soon remarried. His second wife was a wealthy widow named Margaretha, and he used some of her money to set up an alchemical laboratory so that he could search for the Philosopher's Stone. He had studied the Paracelsian doctrine of signatures and speculated that it might be possible to make gold from a substance that had a golden color. He also remembered a piece of alchemical lore according to which the Philosopher's Stone could be obtained from something in the human body. This something, Brandt concluded, could easily be urine.

In 1669 Brandt collected 50 buckets of urine and allowed it to evaporate until it "bred worms." He then boiled the urine to further concentrate it and kept the residue in his cellar until it turned black. Next he distilled the concentrated urine and collected the distillate under water in a flask, obtaining a transparent waxy substance. When this substance was removed from the water, it glowed in the dark and sometimes burst into flame. At first, Brandt thought he had found the Philosopher's Stone. However, further experiments with his phosphorus proved fruitless. By 1675 he had been working with the substance for six years, spending most of his wife's money in the process. Deciding that there was no need to continue to keep the substance he had discovered secret, he showed it to some of his friends and neighbors. He didn't reveal his method for making phosphorus, but did say that it came from the human body.

HERREN KUNCKEL AND KRAFT

Brandt never published an account of his discovery of phosphorus. As a result, during the next century its discovery was erroneously attributed to the German alchemist Johann Kunckel, who was born in 1630. Kunckel's father was the court alchemist of Duke Frederick of Holstein, and Kunckel took up alchemy himself as a young man, holding positions as alchemist in the courts of Duke Franz Carl of Sachsen-Lauenburg and of John George II, the elector of Saxony. Let go from the latter position in 1675, he got another position teaching alchemy at the University of Wittenburg.

It was while he was at Wittenburg that Kunckel began investigating luminescent materials, substances that glowed in the dark after exposure to light. Thus he was intrigued when he heard that a man from Hamburg had found a new substance that glowed more brilliantly than any other. Kunckel visited Brandt, who showed him his phosphorus but refused to describe how he had made it. By this time Brandt was short of money, but he still refused to sell his recipe. Intrigued by what he had seen, Kunckel wrote to the Dresden alchemist Daniel Kraft,* describing Brandt's phosphorus. Kraft immediately journeyed to Hamburg and asked Brandt if he was

*Spelling was not standardized in those days, and the name is variously spelled Kraft, Krafft, and Crafft.

willing to sell any of it. Brandt was willing, being by then desperate for some ready cash. He offered to sell Kraft all that he had and all that he made in the future.

While Brandt and Kraft were talking, there was a knock on the door. It was Kunckel. Brandt stepped outside, saying that his wife was sick and shouldn't be disturbed. Kunckel wanted to buy some of the phosphorus too. But being in the middle of a negotiation, Brandt declined to sell him any, saying that a recent attempt to prepare more had been a failure. He admitted that he had made the phosphorus from urine, however. When Kunckel left, he went back inside to negotiate further with Kraft. A deal was concluded. Brandt was to receive 200 thalers and in return gave Kraft the phosphorus and agreed to keep silent about his method, and in particular not to tell Kunckel how the substance was made.

Kunckel went back to Wittenburg and began a series of experiments with urine. Unsuccessful in his initial attempts, he wrote to Brandt, asking again for information about preparing phosphorus. Brandt would disclose no information, but he said that he had a bargain with Kraft to keep the process secret. But Kunckel persisted and, finally, in July of 1676, he succeeded in preparing phosphorus. Shortly thereafter, he received a letter from Brandt, who was apparently again in need of money, offering to sell his recipe for making phosphorus. Naturally, Kunckel turned down the offer.

Unlike Brandt, Kunckel published a paper describing the properties of phosphorus, without revealing anything about how it was made. He also discovered that phosphorus could be made from many organic substances and claimed that he could make phosphorus from mammals, fish, birds, and plants. Thus, for more than a century, he came to be regarded as the discoverer of phosphorus.

Meanwhile Kraft found a way of turning phosphorus into gold—by giving demonstrations of its properties in the courts of Europe. Ever since Guericke's experiment with the copper hemispheres, scientific demonstrations had been quite fashionable. Kraft claimed that he had discovered the substance and never mentioned Brandt.

Thus, Brandt remained in obscurity, while both Kraft and Kunckel were given the credit for discovering the element that he had first prepared.

LEIBNIZ

In 1677 Kraft demonstrated the properties of phosphorus in Hanover at the court of Duke Johan Frederick of Saxony, where the philosopher Gottfried Wilhelm von Leibniz, who served as librarian to the duke, witnessed it. Leibniz, who had a lifelong interest in alchemy, knew Kraft well. The previous year he had entered into an agreement with Kraft and another alchemist, Georg Hermann Schuler, according to which they would share the profits if any of them succeeded in making gold. Leibniz was intrigued by Kraft's demonstration and asked him how he made phosphorus. Kraft wouldn't say. He couldn't have told Leibniz how phosphorus was made if he wanted to. Only Brandt and Kunckel possessed that secret.

Some time later, on a trip to Hamburg, Leibniz learned that a local resident, a man named Brandt, also knew how to make phosphorus. Leibniz sought him out and learned that he indeed knew how to produce it. Brandt, who was once again short of money, told Leibniz that phosphorus was made from urine and even offered to show Leibniz how to make it if he was paid enough. By now Brandt had decided that Kraft had been exploiting him and was refusing to sell his former business partner any more phosphorus. Leibniz told Brandt that he would talk to his employer. When he returned to Hanover, he suggested to the duke that Brandt be given a position as court alchemist. The duke offered Brandt a salary of 10 thalers a month, which he accepted, but not before considering a similar offer from the ruler of another German principality.

Brandt's method of producing phosphorus was inefficient. It is estimated that he extracted about 1 percent of the phosphorus that was present in urine. Thus very large quantities of urine were needed to obtain even moderate quantities of phosphorus. Because Leibniz

wanted to mass-produce phosphorus, it was necessary to obtain enormous quantities of the raw material from which it was made. At first,
Leibniz got the urine from the soldiers at a nearby army camp. Then
he had an even better idea. One of the duties that he performed for
the duke was to act as a kind of minister of mines. Learning that the
miners in the Harz Mountains consumed even more beer than the
soldiers, he arranged to have their urine shipped to Brandt in large
barrels over 60 miles of roads.

But then the duke sent Leibniz away on a diplomatic mission. By
the time the philosopher returned, he had turned to other endeavors
and lost interest in his phosphorus project. At this point Brandt dis-
appears from history, and all that is known about the rest of his life is
that he lived a fairly long time. Brandt might have still been alive in
1710, when he would have been in his eighties, according to Leibniz.
Leibniz remarked that at least he had not heard of Brandt's death.

Brandt must have produced a number of alchemical manuscripts
during his life, and he certainly wrote letters to, and received letters
from, other alchemists. However, none of these papers has survived.
It is only due to Leibniz that we know that Brandt was the true discoverer of phosphorus.

BOYLE

In 1677 Charles II invited Kraft to England to demonstrate his phosphorus to the royal court. Kraft replied that he would do so for a fee
of a thousand thalers. This was a lot of money,* but Charles
apparently didn't want to forego witnessing a monumental discovery
his continental something had seen, and he agreed. Kraft, after all,
was the only person capable of performing such a feat.

---

*It is difficult to compare the values of currencies in widely separated
eras. However, a thousand thalers is roughly equivalent to a low five-figure
sum in today's dollars.

When Kraft arrived in London, Boyle contacted him and invited him to put on a display for the fellows of the Royal Society also. Kraft agreed, and on a September evening he arrived at Boyle's home, where the fellows had gathered. After the room had been darkened, Kraft passed around a bottle containing a small piece of phosphorus. Boyle wrote later that it glowed "like a cannon bullet taken red hot out of the fire, except that it was more pale and faint." But when the bottle was shaken, Boyle went on, the phosphorus glowed more brightly and emitted flashes of light.

Then Kraft exhibited a tube containing a small amount of phosphorus at one end, which made the whole tube seem to glow. He then took another lump of phosphorus out of its container and allowed the fellows to hold it in their hands. They say that it emitted no smoke or fumes. Kraft then shattered this piece of phosphorus into fragments, which continued to shine after he scattered them on the floor. At this point, fearful that the phosphorus might be burning his sister's Turkish rug, Boyle had candles brought in to light up the room and examined the rug. It was undamaged.

After the candles were taken away, Kraft put some phosphorus on his finger and wrote the word "DOMINI" on a sheet of paper. According to Boyle, the glowing letters seemed a "mixture of strangeness, beauty and frightfulness." Kraft rubbed his finger on the back of Boyle's hand and on the cuff of his coat. These glowed "very vividly." Kraft concluded his performance by attempting to ignite a little gunpowder with the phosphorus. This part of the demonstration failed. However, Kraft promised to return the following week to try again.

The second time Kraft succeeded in igniting the gunpowder with the glowing phosphorus. After the demonstration Boyle asked Kraft to leave a little of the phosphorus with him, or at least tell him how it was made. When Kraft declined, Boyle offered him a secret alchemical formula in return for the recipe. Again Kraft declined. However, he did say that phosphorus was made from something "that belonged to the body of man." After Kraft left, Boyle pondered the matter, and guessed that urine might be the source of phosphorus. He had used urine in some of his own alchemical experiments and had a supply stored in his laboratory.

However, Boyle didn't actually try to make phosphorus until the following year, when he had an assistant boil down large volumes of urine into a paste. But he wasn't sure what the next step should be. Boyle and his assistant tried every procedure for obtaining phosphorus that they could think of, but nothing worked. It then occurred to Boyle that when Kraft said phosphorus was made from something that "belonged to the body of man," he might have been referring not to urine but to excrement. But, however much he and his assistant tried, they couldn't obtain phosphorus this way either. Then, in 1678 or 1679 Boyle hired two German alchemists, a man named Johann Becher and his assistant Ambrose Godfrey Hanckwitz. Neither Becher nor Hanckwitz knew how to obtain phosphorus, but they did know of a man who did: Brandt. When Hanckwitz next visited Germany, he arranged a meeting with Brandt. Brandt declined to give Hanckwitz the recipe, but he gave him an important piece of information: phosphorus would be given off only if the urine residue was heated to very high temperatures.

When Hanckwitz returned to London, he delivered this information to Boyle, and it wasn't long before the two achieved their goal of extracting phosphorus from urine. Boyle immediately began a series of experiments in which he studied the properties of phosphorus. He didn't reveal the method of obtaining phosphorus, but he described his studies of the material's properties in clear detail. Boyle's method of extracting phosphorus was much more efficient than Brandt's. Instead of following alchemical recipes, he treated extraction as a chemical problem and succeeded in getting nearly all of the phosphorus contained in the urine, while Brandt had extracted only about 1 percent of the available element.

Boyle seems to have been fascinated with the new element, which he learned to prepare in both solid and liquid forms. (Phosphorus melts at a temperature only a few degrees above that of the human body and the melting point can be lowered by mixing it with other materials.) He often played little tricks with the material, dipping his finger into liquid phosphorus, as Kraft had done for him, and drawing glowing lines with it on table linen or on the hands of guests at his sister's house. Meanwhile, he continued experimenting, writing paper after paper. In 1682 he published a little book titled *New Experiments and Observations Made upon the Icy Noctiluca*. "Icy noctiluca" was his name for phosphorus.

In 1682 Hanckwitz, who had ceased using his German name and now called himself Ambrose Godfrey, decided to leave Boyle's employ. By this time he had become very proficient at extracting phosphorus, and he decided that he could go into business selling it. With some financial help from Boyle he set up a business which soon began to flourish. Godfrey was able to sell all the phosphorus he could make. He sold it to natural philosophers, to alchemists, to physicians, and to those who wanted to put on shows as Kraft had done. Phosphorus soon acquired a reputation as a medication that could cure almost anything and was reputed to be an aphrodisiac as well. In reality it is poisonous. But in those days the physicians were always looking for new ways to kill their patients, and they began using the substance with enthusiasm. During the 1720s, Godfrey published a series of papers on phosphorus and on acids containing phosphorus. In recognition of his work, he was made a fellow of the Royal Society in 1730 and he died a wealthy man in 1741. The firm that he founded remained in business until 1915.

NEW ELEMENTS

It might seem that not much had been done to advance the science of chemistry. Although a new element had been discovered, it was not recognized as an element, and its main use was to exhibit it as a curiosity. The discovery of phosphorus had impoverished Brandt, while Kraft and Godfrey (né Hanckwitz) became wealthy. Its discovery had done little but affect a few individual lives. However, the discovery of phosphorus was an important event. Boyle's experiments with the substance and his exhaustive descriptions of them provided a model for future chemical research. At the beginning of the seventeenth century, alchemists kept their procedures secret. At the end of the century, it was customary to make what one had learned known. Kunckel, Leibniz, Boyle, and Godfrey all published papers on phosphorus. To be sure, there were still secrets; Godfrey never revealed how his phosphorus was made, but then one would not have expected him to do so. It would endanger his profits. If he kept his recipe secret, he was only doing what Coca-Cola does today.

Most importantly, chemists were beginning to realize that they did not yet know the composition of the universe. If one new chemical substance had been found, why could there not be more? Who could say what discoveries were yet to be made? And they were made. When Brandt first made phosphorus in 1669, no new element had been discovered for more than 150 years. This situation was soon to change, however. Some 15 new elements were to be found during the eighteenth century.

KOBOLDS, OLD NICK'S COPPER, AND FROG GOLD

After the discovery of phosphorus, 66 years passed before another new element, cobalt, was discovered. Cobalt compounds were known since ancient times and had been used to color glass since the sixteenth century. They were collectively known under the name "kobold." Miners believed that the presence of these substances in mines was the work of malicious gnomes called kobolds, who wanted to poison the miners.

During the sixteenth century it was discovered that cobalt compounds could be used to give an intense blue color to glass. But no one knew anything about their composition, or guessed that a metal could be obtained from them. Thus, it is not surprising that cobalt metal was discovered by a glassmaker. In 1735 the Swedish glassmaker Georg Brandt (no relation to Hennig Brandt) examined cobalt ore and succeeded in obtaining a new metal from it. But Brandt was an artisan, not a chemist, and the impure cobalt that he obtained was only a novelty to him. However, chemists later confirmed that it really was a new metal, and the intensive study of cobalt compounds began around the beginning of the nineteenth century.

Nickel has properties like cobalt's, and it was also used to color glass, coloring it green rather than blue. The name "nickel" is derived from German copper miners' name for nickel ore, *kupfernickel* ("Old Nick's copper"). Copper ore and nickel ore look very much alike, and the miners believed that the nickel ore had been planted in mines by the devil to deceive them.

In 1751 the Swedish mineralogist Axel Cronstedt studied kupfer-nickel samples. When he dissolved the mineral in acid he got a green solution. But this solution definitely contained no copper (solutions of copper compounds are green or blue). When an iron object is placed in a copper solution, copper is deposited on it. This didn't happen with a solution of kupfernickel. Realizing that the ore he was dealing with was something other than copper, Cronstedt smelted some of it, obtaining a whitish metal. At first, many chemists refused to accept that Cronstedt had discovered a new element. Many of them believed that nickel was probably a mixture of cobalt, arsenic, and iron. But in 1775 the Swedish chemist Torbern Bergman prepared a sample of nickel that was purer than Cronstedt had obtained and showed that no alloy of cobalt, arsenic, iron, and copper could have the properties that it exhibited.

Platinum is sometimes found in nature in a fairly pure state, so platinum nuggets must have been found in nature in ancient times. But the metal remained unknown to European chemists until the eighteenth century. Once they began to study it, they soon realized that they were dealing not with one new element but with five.

The story begins in 1735, when Don Antonio de Ulloa was one of two officers in charge of a French-Spanish scientific expedition to Peru. While he was in Peru, Ulloa came across some platinum and wrote an account of it. The Spaniards called platinum *platina del Pinto* ("little silver of the Pinto River"), and the mines in South

America were soon using it to adulterate gold. The Spanish government responded by closing the mines and ordered that the platinum,
which then had no commercial value, be thrown into the ocean. The
British did not have a high opinion of the value of the metal either;
they disdainfully called it "frog gold." And of course there was little
market for it. Although it was even rarer than gold, it seemed less
attractive, and it had no apparent uses.

In 1741 the English metallurgist Charles Wood sent a certain
Dr. Brownrigg, an English physician, a specimen of platinum that had
been found in Colombia. Brownrigg called platinum a "semi-metal"
and gave his specimen to the Royal Society in 1750. Chemists soon
became interested in the new metal, and it wasn't long before they
realized that naturally occurring platinum was not a pure metal.
Other elements were alloyed with it in various proportions. This led
in turn to the discovery of other new elements.

In 1803 William Hyde Wollaston, a British physician who became
famous for his research in metallurgy, mineralogy, and optics,
succeeded in extracting a white metal from platinum. He named the
new element palladium, after the asteroid Pallas, which had just been
discovered the previous year. In the same year the English chemist
Smithson Tennant obtained two new metals, which he named iridium
and osmium, from platinum. And in 1828 the Russian chemist Karl
Karlovich Klaus reported that he had obtained three new metals from
platinum mined in the Urals. However, the existence of only one of
them, which Klaus called ruthenium, was confirmed.

GASES

Many metals are relatively inert; they don't combine chemically with
other substances as easily as non-metals do. For example, iron rusts
and copper oxidizes. The green color of copper domes comes from a
thin layer of a copper carbonate. But neither iron nor copper spontaneously combines with elements other than oxygen that are found in
the natural environment. Otherwise, we would not have copper coins,

and we could not make automobiles from steel. Many gaseous elements, on the other hand, combine very readily with a wide variety of other elements. Oxygen, for example, is one of the most abundant elements in the Earth's crust, where it exists in the form of such chemical compounds as oxides and sulfates. Hydrogen is explosive when it comes into contact with oxygen, and chlorine is extremely reactive.

Gases play an important role in chemistry. But at the beginning of the eighteenth century, none of them had yet been discovered. I will describe the discovery of hydrogen, oxygen, and nitrogen in Chapter 5. The identification of these elements was such a significant event that it deserves the greater part of a chapter. For now I will confine myself to saying a few words about the discovery of chlorine.

Carl Scheele was born in 1742 in a part of present-day Germany that was then part of Sweden. Apprenticed to an apothecary at the age of 14, Scheele became fascinated with the properties of the chemical compounds he worked with. After serving his apprenticeship, he became an apothecary's assistant in Stockholm. He must have lived a parsimonious life, because he soon had enough money to buy an apothecary's shop in the town of Köping in central Sweden, where he set up to perform chemical experiments.

Scheele worked obsessively in his laboratory, making one discovery after another and his discoveries attracted attention. He was elected to the Swedish Academy of Sciences at the age of 32, and he later received offers of professorships from many important universities. However, Scheele preferred to remain a provincial apothecary working undisturbed in his laboratory.

One of the notable events in Scheele's life was a discovery that he didn't make. In 1774 he obtained chlorine gas by dissolving a new element that he had discovered, manganese, in hydrochloric acid. He gave the gas the cumbersome name "dephlogisticated* marine acid

---

*The word is a reference to the concept of phlogiston, which will be discussed in Chapter 5.

air" and failed to recognize that it was a new element. For many years thereafter, chlorine was regarded as a compound containing oxygen, and it wasn't until 1810 that the English chemist Humphrey Davy showed that it was an element.

ELECTRICAL CHEMISTRY

Humphrey Davy was born in 1778 at Penzance in Cornwall of middle-class parents. His father died in 1794 and a year later the youth was apprenticed to a surgeon. At the time Davy hoped that he would eventually be able to study medicine. He also had an interest in chemistry, and around this time he began to perform chemical experiments after reading a French textbook* on the subject. As a youth Davy wrote poetry and hoped to eventually publish a book of verse. However, this idea "fled before the truth" once he began the serious study of science.

In 1798, convinced that some of the gases that chemists had recently discovered might prove to be useful in the treatment of tuberculosis, Thomas Beddoes, a former lecturer in chemistry at Oxford University, founded a Pneumatic Institute in Bristol with the intention of carrying out a series of experiments. On the recommendation of a mutual acquaintance, Beddoes hired Davy as a research assistant.

Young men often tend to be foolhardy and Davy, performing a series of risky experiments, was no exception. For example, in order to determine the effects of gases on human beings, he inhaled them. Once he inhaled nitric oxide, even though he knew that nitric acid was formed when the substance came into contact with moist air. Not surprisingly, he burned his mouth and larynx and probably also his lungs. Another time he inhaled "water gas," a combination of

---

*The text was *Elements of Chemistry* by Antoine Lavoisier, who will be discussed in Chapter 6.

hydrogen and carbon monoxide, nearly asphyxiating himself in the process. Yet another time he tried inhaling hydrogen cyanide gas, the gas that was later used to execute people in gas chambers, but fortunately Davy's was not in a very concentrated form.

Davy's research was to make him famous, not because he performed dangerous experiments, but because he discovered a new recreational drug. In 1800 he published a 580-page book titled *Researches, Chemical and Philosophical, Chiefly Concerning Nitrous Oxide, or Dephlogisticated Nitrous Air, and Its Respiration* in which he discussed his discovery of, and researches on, "laughing gas." Describing its effects, Davy wrote:

> A thrilling, extending from the chest to the extremities, was almost immediately produced. I felt a sense of tangible extension highly pleasurable in every limb; my visual impressions were dazzling, and apparently magnified. . . . By degrees, as the pleasurable sensation increased, I lost all connection with external things. . . . I existed in a world of newly connected and newly modified ideas. I theorized; I imagined that I made discoveries.

Davy used nitrous oxide for recreational purposes himself, and he also found it to be a good hangover remedy. He persuaded the poet Samuel Taylor Coleridge (later a great admirer of Davy) and Dr. Peter Roget (the Roget of the Thesaurus) to try laughing gas, and he lectured about its use. Davy suggested that laughing gas be used as an anesthetic, but it was not used for medical purposes until 1846. On the other hand, it was widely used in the student "saturnalia" that were held in chemical laboratories.

But Davy was no nineteenth-century Timothy Leary. He was a brilliant chemist, and his talents were recognized early on. In 1801 he was named professor of chemistry at the new Royal Institution, which had been founded in London only two years earlier. The institution, in spite of its name, had no connection with the British government.

It was a privately funded organization created to publicize the improvements science could make to the quality of life of the middle classes and the poor. However, the king permitted his name to be included in the list of the founders, which was sufficient to make the institution "royal."

One of Davy's first projects was to apply the newly invented electric battery to chemistry. The Italian physicist Alessandro Volta had demonstrated the first one only the previous year, 1800, and Davy immediately saw a use for them in his research. Soon after the first batteries were made, it was discovered that the electrical currents that they produced could be used to decompose chemical compounds. For example, if positively and negatively charged electrodes were inserted in water, oxygen was released at the negative electrode and hydrogen at the positive one. This phenomenon is called electrolysis.

By the beginning of the nineteenth century, caustic soda (sodium hydroxide) and caustic potash (potassium hydroxide) had come to be very widely used in chemical laboratories. Both substances were thought to be elements; no one had ever broken them down into their chemical components. Nevertheless, Davy began to wonder if they might not turn out to be compounds after all, so he decided to subject them to electrolysis to see what happened. Realizing that the more powerful the batteries used, the better the chances of success, he had an assistant connect together all the batteries that the institution possessed. He then dissolved some caustic potash in water and passed an electrical current through the solution.

The experiment was a failure. All that happened was that hydrogen was released at one electrode and oxygen at the other. But Davy was undaunted. If the experiment didn't work with a solution of potash, he could repeat it without the water. However, there was a problem. Cold, dry, caustic potash was an insulator; it didn't conduct an electric current. So Davy tried melted potash instead. Finally in 1807, after a series of failures, he obtained a new metal from the potash, which he named potassium.

Potassium was unlike all the other metals then known. It had a tendency to burn up or explode as soon as it was formed. When thrown into water it bounced around on the surface while making a hissing noise, until one had a solution of caustic potash again. It flamed up in acids and ate into glass. It burned even when placed on ice. Because it combined so readily with other elements it was difficult to study. But Davy soon found a way of preserving samples of the new element. Potassium, and also sodium, which had similar properties and which Davy discovered the same year, could be kept intact in kerosene. Thus, Davy was able to establish that it was indeed a metal, although a very light and reactive one.

Davy used similar methods to isolate four other new metallic elements: calcium, magnesium, barium, and strontium. But again, his first experiments were failures. He had to build new, more powerful batteries before he could achieve success. However, he was unable to break down compounds containing such elements as aluminum and silicon because the technology available at the time was insufficient to allow him to do so. He also made his share of mistakes. He attempted to break down substances that really were elements, such as carbon and nitrogen. At one point, he imagined that he had achieved success and announced in a lecture that sulfur, phosphorus, carbon, and nitrogen were really compounds. Naturally it soon became apparent that he was terribly wrong.

Davy was knighted in 1818. By 1827, his health failing, he emigrated to Rome and lived there, "a ruin among ruins," until his death in 1829. But Davy's contribution to science continued, in a way, long after his death. In 1813 he had the good judgment to hire a former bookbinder's apprentice named Michael Faraday as an assistant. By the time Davy died, Faraday was already beginning to forge a reputation as one of England's great scientists.

DISCOVERY BY FIRE

In the years between 1817 and 1827 chemists discovered six new elements: lithium, selenium, cadmium, silicon, bromine, and aluminum.

Most of these are relatively common and some are common indeed. For example, silicon and aluminum are the second and third most abundant elements in the Earth's crust. The rarest of these elements, selenium, is twice as abundant as silver and 20 times more abundant than gold, and it is relatively easy to obtain because it often occurs in sulfur deposits.

However, chemists had, as yet, no way to detect elements that were present in mineralogical samples only in minute quantities. This became possible only when the German chemist Robert Bunsen and the German physicist Gustav Kirchhoff developed a new method of analyzing chemical samples. Bunsen was not self-taught as Scheele and Davy were, and he never had to serve an apprenticeship. His father, who was a professor at the University of Göttingen, saw to it that he got a good education, and Robert earned the degree of doctor of science at the age of 20. After receiving his degree, the young man spent a year and a half on a tour of Europe during which he met many of Europe's leading chemists. When he returned to Göttingen he was appointed to a minor teaching post at the university.

Bunsen never married, and he soon became one of those bachelors whose life followed a set pattern. He got up every morning at dawn and sat down at his desk to write accounts of his work. He then went to the university to deliver lectures. When these were concluded, he went to work in the laboratory until dinnertime. After dinner, he went for a walk, and then returned to the laboratory again. Bunsen's life followed this pattern when he was 25, and when he was 70 his routine was unchanged.

In 1852 Bunsen became professor of chemistry at the University of Heidelberg. In 1854 a gas works was set up in the city and, realizing how useful gas burners would be in his work, Bunsen had gas piped into his laboratory. When none of the burners he tried was satisfactory, he devised his own. Called the Bunsen burner, it is still seen in all chemical laboratories today. The Bunsen burner produced a steady colorless flame, and the intensity of the flame could be adjusted. It was a far better piece of apparatus than the alcohol and oil lamps that chemists had previously used.

Bunsen soon noticed that if different substances were held in the flame of his burner, different colors were seen. For example, if a glass tube was held in the normally colorless flame, the flame took on a yellow hue. Copper turned the flame green, calcium made it brick red, and sodium produced a bright yellow. He realized that what he had was a flame test that could be used as a tool in chemical analysis, though of course there were serious limitations. For example, if a sample contained a large amount of calcium and a small amount of strontium, the strontium was not detected because its color was masked by the calcium's much brighter color. One day in 1859 Bunsen happened to mention the problem to his friend Gustav Kirchhoff, who also lectured at Heidelberg. Kirchhoff, who was a physicist, immediately saw a solution. He turned up in Bunsen's laboratory with a prism, two small telescopes, and a cigar box with black paper glued to its insides. He set up his apparatus so that the light from one of the telescopes collected the light from a Bunsen burner and focused it on the prism. The prism split the light into the various colors of the spectrum, which could be viewed with the second telescope.

After Kirchhoff set up this device, called a spectroscope, Bunsen placed small quantities of different chemical substances on a platinum wire. Holding the wire in the flame he looked through the telescope's eyepiece. When he put a speck of sodium chloride—ordinary table salt—on the wire he saw that the sodium emitted light at two narrowly separated wavelengths in the yellow part of the spectrum. When he put potassium in the flame he saw two pale violet spectral lines (the apparatus caused colors to be observed in the form of vertical lines) against a dark background. Lithium gave a bright red line and a less intense orange one. Strontium produced a blue line and several red ones. It appeared that each element had its own characteristic signature, which both Bunsen and Kirchhoff quickly learned to recognize.

The beauty of the method was that if a sample contained a large quantity of one element and a small quantity of another, both were detected. If there was, for example, a great deal more calcium than strontium in a sample, the spectral lines produced by the strontium were still seen. If the disparity was too great, the strontium lines might be hard to see, but this could be overcome by using chemical methods to remove some of the calcium.

The collaboration between Bunsen and Kirchhoff continued. Bunsen wanted to look for new elements while Kirchhoff was more interested in using the spectroscope to determine the chemical composition of the sun. However, this caused no conflict. Both Bunsen and Kirchhoff had ample opportunity to pursue their respective lines of research.

Kirchhoff eventually found some 30 different solar elements. Meanwhile, Bunsen systematically examined samples of material from numerous sources. He looked at the spectra of the elements found in minerals, in ores, in water obtained from different sources, and even in animal tissue. At first he saw only the same familiar lines over and over again. And then, one day in 1860, when analyzing some mineral water from the springs at Durkheim, Bunsen observed two blue lines that he had never seen before. He knew that it must be a new element, and named it cesium, after the Latin word *caesius*, which means "sky blue."

Bunsen hadn't actually observed any cesium; he had seen only the light that it emitted. He realized that because it was present in the water only in minute quantities, it could be very difficult to get a sample. If he tried to get cesium in his laboratory it might be years before he succeeded, so he decided to solve the problem by brute force. He commissioned a chemical factory near Heidelberg to spend several weeks evaporating and chemically treating some 12,000 gallons of the Durkheim water.

Meanwhile he experimented further with the water and noticed another set of unfamiliar spectral lines. This had to be a second new element, which Bunsen confidently named rubidium after the Latin word *rubidius*, which means dark red. Realizing that the treatment the mineral water was undergoing at the chemical factory would provide him with a sample of rubidium, too, he simply waited for the results. They turned out to be just what he had expected. The 12,000 gallons of water yielded 10 grams of a rubidium compound and 7 grams of a compound containing cesium. Both cesium and rubidium turned out to be light silvery metals. Their appearance and chemical behavior were similar to those of sodium and potassium and, like the elements that Davy discovered, they had to be kept in kerosene or they spontaneously burned up.

Kirchhoff's and Bunsen's results created a sensation within the scientific community. Soon scientists throughout Europe were using spectroscopes in their research. Inevitably, some of them found yet more new elements. In 1861 the English chemist and physicist William Crookes discovered thallium, a heavy metal, in a sample of clay. In 1863 two German chemists, Ferdinand Reich and Hieronymous Richter, discovered another new metal, which they called indium. And in 1868 the French astronomer Pierre Jannsen observed the sun with a spectroscope during an eclipse and discovered the spectral lines of a new element, helium, in the sun's atmosphere. This was an especially sensational discovery, because helium had never been observed on Earth.

All in all, some 78 new elements were discovered during the eighteenth and nineteenth centuries. In 1700 chemists knew of only a handful of elements, and they had no way to be sure whether they really were elements. By 1899 the list of known elements had grown enormously, raising the inevitable question: Why were there so many?

CHAPTER 5

A Nail for the Coffin

W hen Johann Joachim Becher was born in 1635 to a Lutheran minister, the Thirty Years War had been raging for eight years. Though the chaotic conditions in Germany prevented Becher from getting much of an education, they posed no obstacle to his advancement. When Becher was 26, the elector of Mainz appointed him to be his expert on the management of local industry. Becher also managed to get himself appointed professor of medicine at the University of Mainz. Although he had no formal medical training, he had recently been given a degree in medicine as a wedding present.

Becher subsequently found employment in the courts of the elector of Bavaria and the emperor of Austria, and in 1673 he turned up in Holland with a plan for making sand into gold. The Dutch were skeptical, but Becher gave a dramatic demonstration to a government-appointed commission and persuaded them to approve the sand project. According to him, large quantities of silver were needed to produce the gold, so this was duly provided.

Becher soon left Holland, claiming later that intrigue against him had made him fear for his life. It is uncertain how much of the Dutch government's silver he took with him. In any case, he sailed to England, where he represented himself as an expert on mining. At one point, Prince Rupert, a nephew of Charles I, sent Becher to the mines in Scotland. The sailing of Becher's ship was held up by storms for four weeks, and he used his enforced leisure to write a book titled *Foolish Wisdom and Wise Folly*. In this work, Becher told the story of his life, incorporating outlandish stories along the way. He wrote of a stone that could make people invisible, of a flask that contained words, and of geese that lived in trees and hatched eggs with their feet.

In 1667 Becher published another book, *Physica Subterranea*, in which he expounded a theory that was to profoundly affect chemistry for more than a century. In the book Becher accepted only two of the traditional four elements, earth and water. He then divided earth into three types, so that in effect there were still four elements. He named the three kinds of earth *terra lapida, terra pinguis,* and *terra mercurialis*. The second of these, which he described as an "oily earth," was supposed to be present in all combustible substances and was released when those substances burned.

This looks like yet another alchemical theory, and of course that is exactly what it is. It is little more than a variation on the three-principle theory of Paracelsus. However, in the hands of Becher's pupil, the German physician Georg Ernst Stahl, it became something more. Stahl wrote a number of books between 1703 and 1731 in which he elaborated on Becher's idea. Renaming Becher's *terra pinguis* "phlogiston" (after the Greek *phlogios* meaning "fiery"), Stahl wrote at great length about its role in combustion and in the calcination* (oxidation) of metals. According to Stahl, these were essentially the

---

*Oxygen had not yet been discovered at this time. Thus, to avoid confusion, I use this eighteenth-century term and also the corresponding term "calx" for the oxide of a metal.

same process; both took place when phlogiston was released into the air. And when calxes were heated with charcoal, which was supposedly rich in phlogiston, the metals could be obtained again.

The idea seemed so plausible that European chemists soon generally accepted the phlogiston theory. To be sure, there were some difficulties. When wood was burned, its phlogiston escaped into the air, and the ashes weighed less than the original wood. But calxes weighed *more* than the original metal. It appeared that the release of phlogiston led to weight loss in one case and weight gain in the other. This apparent contradiction seems not to have bothered Stahl. He simply didn't discuss the problem in his books. Other chemists explained the weight gain by postulating that particles of fire were incorporated into a metal when a calx was formed. Yet others suggested that there were two kinds of phlogiston, one with weight, the other with the property of "levity." When the phlogiston containing levity was released, the substance that had contained it became heavier. Some chemists even denied that there was any weight gain. Eighteenth-century chemistry was primarily a matter of observing chemical changes, and measurements were often crude. Furthermore, the chemists of the time commonly used large burning lenses, which were capable of producing high temperatures. If a lens vaporized some of the metal or calx on which it was trained, a decrease in weight might indeed be observed after calcination.

No one had ever isolated phlogiston, and this hypothetical substance apparently behaved in a contradictory manner. Nevertheless, an incorrect theory was quickly accepted, because what chemistry desperately needed then was a theory that could be used to guide research. The phlogiston theory, as we shall see, performed this function admirably.

A more nearly correct theory of calcination and combustion had been proposed in the previous century. In 1630 the Frenchman Jean Rey theorized that the weight increase during calcination came about because air was incorporated into the calx. However, by the time the phlogiston theory was proposed, Rey's idea had been forgotten. It

didn't fit in with the prevailing idea that fire broke substances down
into their components. And of course the phlogiston theory did.

HENRY CAVENDISH

Henry Cavendish was descended from the dukes of Devonshire on
one side and from the dukes of Kent on the other. His father, Lord
Charles Cavendish, was either the third or the fifth son of the second
duke of Devonshire. His mother, Lady Anne Grey, was the daughter
of Henry, duke of Kent. She was not in robust health when she
married, and she died when Henry was only two years old.

Henry was born in 1731. Little is known about his childhood, but
in 1749 he matriculated at Cambridge University, though nothing is
known of his course of study there. However, it must have included
mathematics, because he exhibited great mathematical expertise in
later life. He remained in Cambridge until 1753, when he left without
taking a degree. The reason for this is unknown, but it has been suggested that he might have had scruples about the religious test that
was required of degree candidates, who were required to assert that
they were bona fide members of the Church of England. During his
adult life, Cavendish never attended any church, and he never identified with any religious sect.

Henry's younger brother, Frederick, entered Cambridge two years
after Henry and also left without taking a degree. After Frederick's
studies were completed, the two brothers set off on a grand tour of
the continent. It is not known how long they remained or what places
they visited. However, there is one story about their trip through
Europe. While staying at a French inn they walked by an open door
and looking in, saw a corpse laid out for burial. Neither of the
Cavendish brothers said anything until the next day, when Frederick
asked, "Brother, did you see the corpse?" "Brother, I did," Henry
replied. Nothing more was said about the matter.

After returning to England, Henry lived with his father until Lord
Charles's death in 1783. His father was not well off for a man of his

position and gave Henry an allowance that was quite small by the standards of the day. After being elected to the Royal Society in 1760, Henry dined regularly at the Royal Society Club. His father provided him with five shillings a day for dinner, never anything more.

Lord Charles Cavendish might not have been wealthy, but he was a natural philosopher and experienced experimentalist. Indeed his research on heat, electricity, and magnetism earned him praise from Benjamin Franklin. Henry must have learned a lot from his father, because he, too, became a meticulous experimenter. Some of Henry's experiments in physics and most of his chemical experiments were performed while he was still living under his father's roof.

At the age of 40, Henry inherited a fortune of more than a million pounds, though it is not known which of his relatives the money came from. Although he was now one of the richest men in England, he spent little on himself. The clothing he wore consisted of frayed family hand-me-downs of the style of the previous century. Cavendish didn't entertain, and his dinner each night consisted of a leg of mutton— and nothing else.

However, he permitted himself one luxury: he acquired a large scientific library. Once, when the library was in a state of disorder, Cavendish hired a man who was in difficult financial circumstances to classify and catalog the books. Sometime later, after the job was completed, one of Cavendish's acquaintances mentioned that the librarian was still finding it difficult to make ends meet and hinted that Cavendish might want to help. Without asking how much was needed, Cavendish wrote out a check for £10,000, a small fortune at the time.

He could be generous in other ways, too. When asked for charitable contributions, he invariably tried to find out what the largest contribution was and then equaled it. According to another story, he once attended the christening of a relative and, learning that it was customary to make a gift to the child's nurse, gave her a handful of gold guineas without bothering to count them first.

Though wealthy, Cavendish was indifferent to money. He was the largest holder of bank stock in England, and cash accumulated rapidly in his bank account. One day his banker came to visit and informed Cavendish that he had a balance of £80,000 in his account. Cavendish answered that he did not want to be "plagued" about it, and that if it was any trouble, he would withdraw the money. The banker assured him that this would not be necessary and suggested that some of the money should be invested. "Do so," Cavendish replied, "and don't come back here to plague me about it, or I will remove it." After receiving his inheritance, he acquired a house in Clapham, on the outskirts of London, where he lived as a recluse, filling the house with scientific equipment.

Cavendish had a high squeaky voice and a stutter, and he avoided speaking to people whenever he could. Throughout his life he was terrified of women, and he sometimes covered his eyes and fled when he encountered them in the street. Nevertheless, he regularly attended meetings of the Royal Society. His devotion to science was total. He seems not to have said much at the meetings, however. As his fellow scientist Lord Brougham observed, Cavendish "probably uttered fewer words in the course of his life than any man who ever lived to fourscore years, not at all excepting the monks of La Trappe."

Because Cavendish was so reclusive, many of the important details about his life remain unknown, and his biographers have generally focused on his scientific work. However, there are numerous stories about his eccentricities. For example, he avoided conversation with his housekeeper and communicated with her by leaving notes. Once when he was at the home of Joseph Banks, president of the Royal Society, a foreign visitor appeared. He had come to London expressly to meet Cavendish, whom he considered to be one of the greatest natural philosophers of his time. At his first opportunity, Cavendish fled and had himself driven home in his carriage.

Cavendish died on February 24, 1810. The several accounts of his last hours are similar, but differ in certain details. According to one story, when he realized that he was dying, he rang for his valet and gave him the following instructions: "Mind what I say, I am going to die. When I am dead, but not before then, go to Lord George

Cavendish and tell him of the event. Go!" Half an hour later Cavendish summoned the servant again and had him repeat the instructions. When the valet returned to the bedroom for the third time, Cavendish was dead.

According to a second account, when Cavendish's servant discovered that his master was dying, he rushed to the house of Sir Everard Home, a well-known physician. Home accompanied the servant to Clapham, where he found Cavendish dying. When he saw Home, Cavendish raged at the servant, saying that at his age any attempts to postpone death would only prolong his miseries. Home nevertheless remained throughout the night, and Cavendish died during the early morning hours.

According to a third story, when Cavendish was dying, he instructed his servant to leave him alone and not to return until a certain later time, when he expected to be dead. However, the servant, who was anxious about his master's condition, returned before the specified time to look in on the dying man. Cavendish, who was still conscious, angrily ordered him out of the room. When the servant returned at the time that his master had stated, Cavendish was dead.

In his will, Cavendish left none of his money to science. Most likely he believed that, because the money had come from his family, it should go back to the family. The famous Cavendish laboratory at Cambridge was founded on the bequest of a relative 61 years after Henry Cavendish's death. He was buried in All Saint's Church in Derby, now Derby Cathedral, but there is no plaque in the cathedral to indicate his interment there. Most likely Cavendish would have been pleased to know that he was to be as anonymous in death as he was reclusive in life.

THE "DISCOVERY" OF PHLOGISTON

In 1766 the Royal Society published Cavendish's *Three Papers Containing Experiments on Factitious Air*, describing his experiments with hydrogen, which is produced when metals are dissolved in acids.

Hydrogen had previously been observed by Boyle and by Cavendish's contemporary Joseph Priestley. However, Cavendish is credited with the discovery of the gas because he was the first to perform experiments for the purpose of determining its properties.

His name for hydrogen was "inflammable air." However, he had no doubt that this gas he had discovered was phlogiston. In science, theory often determines what we observe. Cavendish was simply interpreting his results in terms of the accepted theory of his day. He was far from the only scientist who did so. For example, when Priestley discovered oxygen, he named it "dephlogistated air."

Cavendish began his studies of the properties of hydrogen by mixing it with air and bringing about an explosion. He soon concluded that "this air, like other inflammable substances, cannot burn without the assistance of common air." He also determined that air was 20.8 percent oxygen, which is remarkably close to the modern figure of 20.9 percent. Next, he performed experiments to determine the density of the hydrogen and concluded that the gas was "5490 times lighter than water or near 7 lighter than common air." He then performed some further, more accurate experiments and corrected these figures to 8,760 and 11, respectively. The modern figure for the latter is 14.4. However, Cavendish's apparatus was quite crude by modern standards, and he must have conducted the experiments very carefully to get so close.

Cavendish's papers next described experiments with carbon dioxide, which he called "fixed air." He studied its solubility in water and its efficacy in extinguishing flames. He concluded that when one part carbon dioxide was mixed with eight parts of ordinary air, candles would not burn. He also determined the density of carbon dioxide, finding that it was about 50 percent heavier than air. This result is astonishingly close to the modern value of 52 percent.

I will pass over the other experiments described in these papers except to say that they were numerous. Cavendish's next paper on chemistry was of much greater interest. Published in 1784 under the title *Experiments on Air*, it described his discovery that water could be made by combining hydrogen and oxygen. After Priestley discovered "dephlogisticated air" (oxygen), Cavendish began to experiment with the gas. In one celebrated experiment he mixed air and hydrogen in a long glass cylinder and caused the hydrogen to burn. He found that the water that condensed in the cylinder had no taste or smell and that no sediment was formed when it was evaporated. He had produced pure water. This was followed by experiments in which hydrogen and oxygen were mixed in a large glass globe and exploded with an electric spark. Again, water was produced. Cavendish also obtained quantitative results, finding that water was a combination of two volumes of hydrogen and one volume of oxygen.

Cavendish didn't realize that he had shown that water was a compound of hydrogen and oxygen and therefore not an element. He was hobbled by the phlogiston theory, the only theory that chemistry then had. He concluded that hydrogen was water saturated with phlogiston and that oxygen was water from which all the phlogiston was removed. When they were combined, of course water was produced. Though his conclusions were erroneous, Cavendish had performed a crucial experiment that put another small nail into the coffin of the still widely accepted four-element theory. Although he misinterpreted his results, he had performed one of the crucial experiments in the history of chemistry.

WEIGHING THE EARTH

Cavendish's most famous experiment, performed when he was nearly 70, is often described as "weighing the Earth." This description is a little misleading, because what he was actually trying to do was determine the Earth's density. In order to do this, he needed to calculate the Earth's mass, though that wasn't his primary goal.

Before Cavendish's experiment, no one knew the strength of the force of gravity.* This might sound a little surprising at first, but not

---

*Perhaps I should add, for the benefit of readers who know a little physics, that Cavendish wanted to measure Newton's gravitational constant.

when you consider that the attraction the Earth exerts on terrestrial objects depends on two things: the mass of the Earth and the intrinsic strength of the force of gravity. If the Earth had twice the mass that it does and gravity were half as strong, we would all have exactly the same weight. Conversely, if the Earth were half as dense and gravity twice as strong, the weight of terrestrial objects would also be unchanged.

In order to make this measurement, Cavendish used an apparatus known as a torsion balance, which consists of a light rod to which two lead balls are attached, one at either end. The rod is suspended on a long slender wire, and two larger and heavier lead balls are placed in stationary positions near the ends of the rod. Knowing that the gravitational attraction between the lead balls was extremely small, Cavendish knew that he had to eliminate any effects, such as air currents, that might interfere with his measurements. So he placed the apparatus in a mahogany case and took some additional precautions, which he described as follows: "I resolved to place the apparatus in a room which would remain constantly shut, and to observe the motion of the arm [the rod to which the smaller balls had been attached] from without by means of a telescope: and to suspend the leaden weights in such manner, that I could move them without entering the room."

If a slight horizontal motion were imparted to the balls, then there would be two forces acting on the rod: the gravitational attraction between the smaller and heavier balls, and a force that was due to the twisting of the wire. These two forces would cause the balls to swing back and forth horizontally in a manner that somewhat resembled the motion of a pendulum. By observing the motion of the rod, Cavendish was able to calculate the tiny gravitational attraction between the balls on the rod and the larger stationary ones. From this he could determine what the intrinsic strength of the force of gravity was, and once he knew this, he could calculate the mass of the Earth and its density.

Cavendish repeated the experiment 29 times and found that the Earth weighed 6×10^{21} (6 followed by 21 zeros) metric tons. The average density of the Earth was 5.48 times greater than that of water. This was a great improvement on the best previous result (obtained by a different method) of 4.5, and it was better than any determination of the Earth's density that was made before the twentieth century. Cavendish's result differs from the modern figure of 5.52 by less than 1 percent. It is much better than a result of 5.67 that was obtained in 1841 from an experiment that was repeated, not 29, but more than 2,000 times.

When Cavendish died, his contemporary and biographer George Wilson wrote that "he did not love; he did not hate; he did not hope; he did not fear." Wilson also described Cavendish as consisting of "an intellectual head thinking, a pair of wonderfully acute eyes observing, and a pair of very skillful hands experimenting or recording." And of course the latter assessment is correct. Cavendish was the greatest experimental scientist of the eighteenth century.

JOSEPH PRIESTLEY

Joseph Priestley, who was born in 1733, in Fieldhead in the county of Yorkshire, was the son of a dresser, a craftsman who treated woolen cloth to give it an even texture. Joseph's father, Jonas, was a Calvinist dissenter. In those days a dissenter was anyone in England who belonged to a religious denomination other than the Anglican Church. Dissenters included Roman Catholics, Jews, and Quakers as well as dissenting Protestants. Technically, being a dissenter was illegal; there were still laws that specified punishments for anyone who did not subscribe to the Thirty-nine Articles of the Church of England. The laws against Roman Catholicism were especially harsh. The saying of mass by a foreigner was a felony; and if an English priest said mass, he was committing high treason. But these laws, which had been enacted in the previous century, were not enforced.

Priestley lived with his parents until he was four years old. He was then sent to live with his grandfather, who had a farm a few miles away, and remained there until his mother died in childbirth in 1739.

When he was nine, he was sent to live with his father's older sister and her husband. Priestley's aunt saw to his education, sending him to several different local schools.

In 1746 he contracted a serious illness, probably tuberculosis. Cared for by his aunt, he slowly recovered, although he remained in frail health for several years. At the age of 15 he left the school he had been attending and began to learn Hebrew at a school operated by a dissenting minister. Finally, in 1752 he enrolled in a dissenting academy in Daventry in Northhamptonshire. At the time it was necessary to be an Anglican to attend universities like Oxford and Cambridge, so the dissenting academies were created to serve the needs of non-Anglican students. These academies provided the best education available in England at the time. Because they were not hobbled by tradition, they were more progressive than the old English universities.

While he was in Daventry, Priestley decided to prepare for the ministry. He had previously studied French, German, Italian, and Hebrew and now began studying Greek. He also began to write manuscripts that were later published as books. He wrote in shorthand, so that he could write them quite rapidly. It was sometimes said later that he could write books faster than other people could read them.

In 1775 Priestley left Daventry to become an assistant minister in a church in Needham Market, Suffolk. By this time, he had repudiated Calvinism and was developing theological ideas of his own, some of them quite unorthodox by the standards of the day. Later, he would describe himself as a "furious freethinker" during this period of his life. This might not have caused any problems if he had not chosen to give a series of lectures based on the material he had written in Daventry. But give them he did, with the result that his congregation dwindled considerably. In turn, this reduced his income, which had not been very large to begin with. Hoping to supplement it, he announced that he would open a school. This idea ended in utter failure. He didn't get a single pupil, which was hardly very surprising, considering the reputation he had gained in the local community.

Priestley left Needham in 1758 to become minister to a congre-

gation in Nantwich in Cheshire. Again he decided to supplement his
income by opening a school. This time he was more successful and
soon had 36 pupils. Until then he read scientific books but couldn't
perform experiments, because he couldn't afford apparatus. But now,
although his teaching duties prevented him from doing much scientific work, his income was enough to buy some scientific instruments.
He simply didn't have the time to minister to a congregation, teach,
and perform experiments too.

LONDON

In December 1765 Priestley journeyed to London, hoping to make
the acquaintance of some of his fellow British scientists. By this time,
he had attained some prominence as the author of a number of text-
books that were based on the courses he taught at his school. While in
London, he performed his first experiments, some of them under the
direction of Benjamin Franklin, who was in London at the time as a
representative of the government of Pennsylvania.

Priestley continued experimenting, and soon he had written a
manuscript titled *The History and Present State of Electricity, with
Original Experiments*. In June of 1766 he was elected to the Royal
Society, and his book on electricity was published in 1767. At this
time, his experiments were primarily in physics. He had not yet
developed the great interest in chemistry that was eventually to lead
him to some very significant discoveries.

LEEDS

In 1767 he accepted a call to become minister to a congregation in
Leeds. During his six years there, he published voluminously on theological, political, and scientific topics. His publications ranged from
pamphlets on a variety of topics to three-volume works. By now,
Priestley had been a Unitarian for some time, and his theological and
political works often became quite controversial. He spoke, for ex-

ample, of the "idolatrous worship of Jesus Christ." Priestley believed that Jesus had been only a man, and he did not hesitate to say so.

His political opinions, too, were quite radical by the standards of the day. He believed, for example, that under certain circumstances revolution was justified, and he attacked the intolerance that made all religious denominations except the Anglican Church illegal. "Let all the friends of liberty and human nature join to free the minds of men from the shackles of narrow and impolitic laws," he wrote, going on to say, "Let us free ourselves, and leave the blessings of freedom to our posterity." Such ideas wouldn't sound very inflammatory today. However, voicing them in eighteenth-century England was a different matter. His ideas were attacked by the archdeacon of Winchester and by the jurist William Blackstone. Priestley answered both men and soon found himself the center of controversy.

He did not neglect science while he was in Leeds, but continued performing electrical experiments, as well as studies of light and of optical and astronomical instruments. His writings on these subjects brought an invitation from Captain James Cook to join Cook's second voyage as the expedition's astronomer. Priestley accepted but some clergymen on the Board of Longitude, which was sponsoring the voyage, blocked his appointment. Opposition to his ideas was widespread at this time. King George III expressed his disapproval of Priestley, and so did many people of lesser rank.

It was around this time that Priestley discovered soda water. "Fixed air" (carbon dioxide) had long been known to chemists, and Priestley had experimented with it a little. One day he had the idea of trying to dissolve the carbon dioxide in water. He succeeded and found that the water fizzed. Priestley gave some of the soda water to friends and then went on to other kinds of research. Some years later the British Navy expressed interest in the use of Priestley's sparkling water as a remedy for scurvy, but naturally it was unsuccessful. However, soda water quickly became popular in other circles, even earning praise from Lord Byron, who wrote the following stanza on the back of the manuscript of his poem *Don Juan*:

I would to Heaven that I were so much clay,
As I am blood, bone, marrow, passion feeling—
Because at least the past were passed away,
And for the future—(but I write this reeling,
Having got drunk exceedingly today,
So that I seem to stand upon the ceiling)
I say—the future is a serious matter—
And so—for God's sake—hock and soda-water!

THE DISCOVERY OF OXYGEN

In 1772 Priestley published an account of five years of experiments with "airs" (gases). The work he described was so important that it immediately established him as one of the great chemists of the day. While he was in Leeds, Priestley discovered three gaseous oxides of nitrogen, including nitrous oxide ("laughing gas") and hydrogen chloride gas. Before he began his experiments, chemists had known of only three gases: hydrogen, carbon dioxide, and air.

He continued his experiments and discovered even more new gases: sulfur dioxide, silicon fluoride, ammonia gas, and nitrogen. However, his most important discovery was oxygen. In June 1774 Priestley got a burning lens with a diameter of 12 inches and immediately began to experiment with it. In one experiment he turned the lens on mercury calx (mercuric oxide) and obtained an "air" in which candles burned more brightly than they did in ordinary air. At first he did not know what to make of this result, so he continued experimenting. He soon found that he could get the same gas from certain other materials, such as lead oxide.

To find out what this "air" was, he performed further experiments. He did tests to see if it resembled nitrous oxide. It didn't. Still puzzled, Priestley turned to other work, including his experiments with sulfur dioxide, and then he had an idea. If nitrous oxide was mixed with air, the quantity of air was diminished (because another oxide of nitrogen was formed, using up some of the oxygen). He tried mixing the nitrous oxide with his new air and found that it was also diminished. Furthermore, the decrease was greater than that which he observed when he mixed nitrous oxide and ordinary air together.

Next, Priestley took a mouse and put it in a container full of the new air, expecting it to live for about 15 minutes. Instead, it was still alive after half an hour. Apparently this gas had something in common with ordinary air, with the difference that it was somehow "better." Priestley continued experimenting, and even he tried breathing some of the oxygen himself. He found that the feeling of breathing it was not much different from that of breathing ordinary air. However, he wrote, "I fancied that my breast felt particularly light and easy for some time afterwards."

Priestley called the substance he had discovered "dephlogisticated air." According to the theory of the day, air was necessary for combustion because there had to be something that would take up the phlogiston that the burning object released. If objects burned better in the new "air," then this meant that it was air devoid of phlogiston. Ordinary air did contain phlogiston, and this caused com-bustion to proceed more slowly. Similarly, nitrogen, with which Priestley also experimented, was "phlogisticated" air. The large quan-tities of phlogiston that it contained prevented it from supporting combustion.

PRIESTLEY'S PATRONS

Priestley wasn't wealthy like Boyle and Cavendish so he didn't have the leisure to spend as much time on scientific experiments as they did. However, by this time his fame had grown and he soon had a patron. William Petty, the second earl of Shelburne, admired Priestley's scientific work and offered him a post supervising the education of his two sons and collecting material on subjects under discussion in parliament. The salary was to be two and a half times what he was then earning, and Shelburne also provided his new employee with a townhouse near the earl's London residence and a house on his estate in Wiltshire. Priestley took up the post in 1773. When he left it in 1780, Shelburne continued to pay him half the £300 salary he had been receiving, and other patrons contributed smaller amounts to make up the difference.

One of the patrons found him a house on the outskirts of Birmingham, where he could live comfortably and devote himself to scientific experimentation. Priestley set up a scientific laboratory in the house. While he lived there some of his patrons died or stopped contributing, but there were always others willing to take their places. Dr. Erasmus Darwin and the pottery designer and manufacturer Josiah Wedgwood, the two grandfathers of Charles Darwin, were among those who contributed to his support.

Priestley did not live in Birmingham in utter tranquility. After the French Revolution began in 1789, many people in England began to fear that there might be attempts at revolution there, too. Some believed that Priestley, who had published writings expressing sympathy with the French republicans, might try to foment it. He was attacked by the press and denounced in the House of Commons. In 1791 rioters burned down his house and destroyed his laboratory, some distance from the house. He found a new house and began to construct a new laboratory. However conditions in England were rapidly becoming intolerable. When the French executed Louis XVI in 1793, the hysteria increased further. By now some of Priestley's scientific colleagues were snubbing him, and his friends were urging him to leave England, saying that it wouldn't be safe to remain. Priestley took their advice, and in 1794 he and his wife sailed to America.

By now Priestley was famous, and he got a warm welcome in the young republic. After he settled at Northumberland, in Pennsylvania, the trustees of the University of Pennsylvania voted unanimously to offer him the chair of chemistry. But Priestley declined the offer. By the summer of 1795 he had set up yet another laboratory and was again busy writing and performing experiments. But then, in 1797, attacks against him began to appear in the press, just as they had in England. The press had been somewhat hostile since his arrival in America and now the attacks intensified. Priestley was characterized as an enemy of religion and of law and order. Priestley had never commented on American politics, but this was conveniently ignored. One journalist even attacked Benjamin Franklin, who had died at the beginning of the decade, because Franklin had been one of Priestley's friends. Although Priestley was often attacked in print, he never had to fear for his life or his property in America, as he had in England. At one point, he contemplated emigrating to France, but soon gave up the idea. He spent the remaining years of his life living in Northumberland, writing, experimenting, and sending papers to the American Philosophical Society. He also arranged for the publication of religious works that had been written years before and left unpublished for various reasons. Priestley's *History of the Corruptions of Christianity* was published in 1797, and his *Index to the Bible* appeared a few months after his death in 1803.

CHAPTER 6

"Only an Instant to Cut Off That Head"

Antoine Lavoisier was the greatest chemist of the eighteenth century. Yet he found no new elements, and he made no discoveries of the magnitude of Cavendish's and Priestley's. Indeed, some of his greatest achievements were negative ones. It was Lavoisier's experiments that once and for all discredited the four-element theory. And it was Lavoisier who almost single-handedly overturned the phlogiston theory. He had the best laboratory apparatus money could buy, and his experiments were models of precision. However, it was his theoretical work that had the greatest impact. Lavoisier discovered a theory of combustion that was far superior to the phlogiston theory. He theorized about the nature of acids and gasses, and he created tables of the elements. Lavoisier sought to modernize chemistry by giving it a new theoretical foundation, and though some of his ideas were mistaken, on the whole he was enormously successful.

Lavoisier was born in Paris in 1743, into a wealthy bourgeois family. At the age of 11 he inherited 45,600 livres (roughly equivalent to $2 million, according to one estimate) from his maternal great-grandfather. And when his grandmother died in 1768, he inherited yet another fortune. As an adolescent he attended the Collège Mazarin, the wealthiest and most prestigious of the Parisian secondary schools. Then in 1761, following a family tradition, he studied law for two years and gained admission to the Order of Barristers in 1763. The barristers were lawyers who had the right to plead cases before the Parlement de Paris, the highest court in France. Barristers enjoyed considerable prestige, and about half of the members of the order did not bother to practice law. They were content with rising to a social position not far below that enjoyed by members of the French aristocracy.

THE ACADEMY

Lavoisier was one of those barristers who never pleaded a case; he became a barrister to please his father. By the time his law studies were concluded, his interests were already primarily scientific. Impatient to make a reputation for himself, he immediately set his sights on gaining admission to the Academy of Sciences, roughly equivalent to the English Royal Society but with quite different admission standards. Gaining membership in the Royal Society was relatively easy. The Academy, on the other hand, had a fixed number of members, and a vacancy occurred only when one of the members died.

According to rules adopted in 1699, when a vacancy occurred, a list of candidates was drawn up and the academicians voted by secret ballot to pare the list down to two or three. This short list was sent to the king, who made the final decision. Lavoisier realized that gaining admission depended on impressing the members of the chemistry section, who would draw up the initial list of candidates. He began his campaign by writing an essay on street lighting.

In 1764 the Academy had announced a competition on the subject of how city streets could best be illuminated, offering a prize of 1,000 livres. By the time the deadline came the following year, no suitable proposals had been submitted. So the Academy doubled the prize. Lavoisier now set about investigating lighting methods in earnest. To sensitize his eyes, he covered the walls of his study with black cloth and lived in darkness for six weeks. There he performed experiments on every kind of street lamp he could think of and compared different types of fuel, concluding that olive oil would be the best. Finally, he wrote a 70-page essay and submitted it.

In the end, the Academy divided the prize between three people, one of whom was Lavoisier, who was presented with a medal on August 9, 1766. The presentation was described by the *Avant-Coureur* (which misspelled Lavoisier's name) as follows: "There was a paper full of curious research and the best physics, done by M. Ravoisier [sic] whom the Academy of Sciences praised. The king awarded him a gold medal, which was publicly presented by the president of the Academy."

At this point, Lavoisier was primarily interested in doing research that would impress the academicians. He read his first paper to the Academy as a "visiting scientist" (that is, a non-member) on February 27, 1765. It dealt with a topic that had been pursued by some of the chemists in the Academy and was titled "The Analysis of Gypsum." Gypsum was the mineral from which plaster was made, so it was a topic with practical applications. The two referees appointed by the Academy to judge the paper reported favorably on it and recommended it for inclusion in the Academy's *Savants étrangers* collection. It is probable that some sort of behind-the-scenes deal had been made because both referees were friends of Lavoisier's father.

In 1766 Lavoisier presented another paper on gypsum. Shortly afterward, a vacancy opened in the chemistry section of the Academy, and Lavoisier proposed himself as a candidate. A list of eight candidates was soon drawn up, Lavoisier among them. However, when the members of the Academy voted, and the list of eight was narrowed down to two, Lavoisier lost out.

In 1768 another academician in the chemistry section died, and Lavoisier again began maneuvering. Once again he presented papers on topics that were known to be of interest to the chemists in the Academy. And once again proposed himself for membership. This time, when the vote was taken, he was among the final two candidates. However, his attempt to gain election failed once more, although he had received more votes in the Academy than his rival, the chemist Gabriel Jars. The king, when choosing between the two, favored Jars, who was older than the 25-year-old Lavoisier and had a record of public service. Nevertheless, the academicians were impressed with the quality of Lavoisier's scientific work and appointed him a supernumerary adjunct in the chemistry section. On June 1, 1768, both Jars and Lavoisier were made members of the Academy.

TAX FARMER

In 1768 Lavoisier became a tax farmer as well as an academician. His grandmother died that year, and he had to decide how to invest his inheritance. An acquaintance suggested that he enter the *Ferme générale* (General Farm), a private company that contracted with the French government to collect many of its taxes. The company had little to do with agriculture. The word "ferme" had a meaning similar to that of the English word "farm" in the expression "farm out." The Farm had 60 stockholders, and Lavoisier purchased a third of a share for 520,000 livres (approximately $20 million). He made a down payment of 68,000 livres and borrowed the remainder. In doing so, he entered an unpopular profession. The Farm was in charge of collecting indirect taxes, such as a salt, tobacco, and alcohol taxes, customs duties, and taxes on merchandise entering Paris. The French populace found these taxes burdensome. They evaded them whenever possible, and the tax collectors were despised. The French government enforced the taxes with harsh measures. In one year alone, more than 10,000 people were arrested for smuggling, that is, surreptitiously transporting salt without paying the tax. One-third of the French penal population consisted of convicts who had been sentenced for smuggling of one kind or another. Lavoisier had made a lucrative

investment, however. It has been estimated that his income from the
Farm, after the interest on his loan, bribes, and other expenses were
deducted, amounted to the equivalent of several million dollars per
year.

The jobs at the Farm were divided among the members. Initially
Lavoisier was a regional inspector for the tobacco commission, working under the supervision of a senior partner named Jacques Paulze.
In 1771 Paulze discovered that he had a problem. His daughter Marie
Anne was 13 years old and was about to leave the convent in which
she had been raised. The Baroness de la Garde heard of this and
decided that Paulze's daughter would make a good wife for her
50-year-old brother, the count of Amerval. Amerval, who was
financially strapped, wanted to make a good marriage and the Paulze
family had money. So the Baroness applied pressure on the abbé
Joseph Marie Terray, who, as controller-general of finances, supervised the General Farm, to persuade Paulze to agree to the match.

Marie Anne was fully aware that, sooner or later, a marriage
would be arranged for her. However, she had no desire to become the
wife of Amerval, whom she considered a fool and whom she
described as "*un espèce d'ogre*" (a kind of ogre). The situation was
serious. Paulze didn't want Amerval as a son-in-law. However, Terray
was threatening to remove him from his post at the Farm. Then it
occurred to Paulze that there might be a way out. If he could quickly
arrange a marriage for Marie Anne with someone else, he would
escape the pressure from Terray. To be sure, Terray might become
vindictive after he heard what had been done, but that was a chance
that Paulze was willing to take. He suggested to the 28-year-old
Lavoisier that he marry his daughter instead.

Lavoisier, though he had not been courting Marie Anne, agreed.
The arrangement was made in November 1768, and the couple was
married in December. It is not known how Lavoisier felt about the
arrangement or even whether he felt attracted to the girl. He was not
a man who confided his feelings to others. As the American historian

of science Charles C. Gillespie wrote, "It does not appear that Lavoisier was capable of intimacy. He wished to be right, not to be known, let alone to be liked." However, it is not unreasonable to assume that Lavoisier was conscious of the advantages of marrying the boss's daughter.

The benefits were immediate. Marie Anne was given a dowry of 80,000 livres, 21,000 to be paid immediately, the remainder over a period of six years. Receipt of the dowry, together with an advance on the inheritance due from his father when he died, allowed Lavoisier to increase his stake in the Farm from a one-third to a one-half share. Terray made no attempt at revenge. After all, he had lost nothing; he hadn't been the one who wanted to marry Marie Anne. He even arranged to have the couple married in his private chapel. And a reception attended by some 200 guests was held in the mansion of Terray's brother.

It turned out to be a harmonious marriage. The newlyweds developed a great affection for one another. Soon after the marriage Marie Anne began to take an interest in her husband's scientific work, and before long she became his collaborator. Lavoisier arranged to have the chemist Jean-Baptiste Bucquet tutor Marie Anne in chemistry, and she learned English so that she could translate English papers on chemistry for her husband. Marie Anne studied drawing with the painter Jacques Louis David, enabling her to make illustrations to accompany Lavoisier's work. Marie Anne also assisted her husband in performing his experiments, and she was always the one who wrote them up.

Without his wife's help, Lavoisier might not have been able to perform so many important experiments, because administrative activities began to devour much of his time. He soon became a senior partner in the Farm and served on several of its committees. He was becoming more active in the Academy of Sciences, serving on committees there also.

PHLOGISTON

Within a few years of his marriage, Lavoisier began to have doubts about the phlogiston theory. He wasn't the first to have them or to have performed experiments to test the theory. During the mid-eighteenth century, the Russian poet and scientist Mikhail Vasilevich Lomosov had heated metals in sealed vessels and found that there was no increase in weight. He correctly concluded that something in the air combined with the metal. The weight gain in calcination was equal to the weight loss when something was removed from the air. Thus there was no need to invoke phlogiston to explain the results. Unfortunately, Lomosov's results remained unknown in Western Europe, where little attention was paid to Russian science.

In 1772 Lavoisier read an article on phlogiston by the French lawyer and chemist Louis-Bernard Guyton de Morveau. Guyton had carried out careful experiments that showed that metals do indeed increase in weight when they are heated in air. Adhering to the then-orthodox theory that calcination involved a loss of phlogiston, he concluded that phlogiston was so light a substance that it buoyed up substances that contained it. Thus the metals were lighter than the calxes. Lavoisier was skeptical. He suspected that the gain in weight was more likely to have been caused by the incorporation of air into the metals.

In October 1772 Lavoisier performed some experiments using a large burning lens owned by the Academy of Sciences. He found that when litharge (an oxide of lead) was heated with charcoal, large quantities of "air" were released. Of course it wasn't air at all—it was carbon dioxide. At the time, Lavoisier was unaware that carbon dioxide—or "fixed air" as it was then called—has properties very different from ordinary air. He was also unaware that Priestley had experimented with numerous different "airs" and had shown that atmospheric air has more than one component. Lavoisier also performed experiments that showed that sulfur and phosphorus also gain weight when they are burned. Marie Anne wrote up the results and Lavoisier promptly deposited a paper on his experiments in the archives of the Academy. This was a common practice at the time. To establish their priority in their discoveries, members of the Academy often gave the secretary their papers in sealed and dated envelopes in the presence of witnesses. Or they had the secretary initial the pages of their papers as they were written.

Lavoisier concluded that ordinary atmospheric air was responsible for combustion and for the weight gain of the substances that were burned. At the time, he was unaware of the experimental work that had been done, most notably by Priestley, on the different kinds of "airs" that were the products of chemical reactions. Nor did he know that it had been demonstrated as long ago as 1756, by the Scottish chemist Joseph Black, that "fixed air" had properties that were very different from those of the air that constituted the atmosphere.

Lavoisier was aware of his ignorance, however, and in 1773 he began an intensive study of the history of chemistry, paying special attention to experiments with the different airs and repeating many of these experiments with new safeguards. However, this only led him to a new error. He now became convinced that fixed air, or carbon dioxide, was responsible for combustion.

He soon discovered that he had good reasons to change his mind. In October 1774 Joseph Priestley visited Paris with his patron Lord Shelburne. During a dinner at Lavoisier's house, Priestley spoke of a new air that he had discovered, obtained by heating mercury calx without using charcoal, that supported combustion much better than ordinary air. Lavoisier performed his own experiments with mercuric oxide, and in April 1775 he read a paper at the Academy in which he identified the gas that supported combustion as "pure air" rather than any particular constituent of air. Lavoisier still hadn't gotten it quite right. However, at the end of 1775 Priestley published a book in which he described his experiments with oxygen. Reading this book and performing further experiments put Lavoisier on the right track. He realized that combustion was supported by a constituent of air, which he called oxygen.

"Oxygen," which is derived from the Greek, means "acid former." Lavoisier had observed that, when non-metals were burned and the combustion product was dissolved in water, an acid was created. For example, burning sulfur produces sulfur dioxide gas, and when this gas is dissolved in water, sulfuric acid is created. Thus he concluded that all acids contained oxygen. Here he was mistaken again, because there are exceptions. The most notable is hydrochloric acid, which is hydrogen chloride gas dissolved in water. The chemical formula of hydrochloric acid is HCL; no oxygen is present.

It might appear that Lavoisier was forever getting things wrong, but this really wasn't the case. As he attempted to understand combustion, his ideas took many twists and turns, until he finally arrived at a correct theory. This is often the way that scientific discoveries are made. Many great scientists, probably most of them, entertain incorrect ideas before finding their way to the discoveries for which they are known.

Lavoisier was the first to identify oxygen as an element. Recall that to Priestley, oxygen was dephlogisticated air, while Cavendish believed that oxygen was water from which all of the phlogiston had been removed. Lavoisier's doubts about the phlogiston theory enabled him to see more clearly than his two English contemporaries and to understand what was really happening when substances burned or when metallic calxes were formed. The element oxygen was combining with other substances.

WATER IS NOT AN ELEMENT

When Lavoisier's career as a chemist began, the four-element theory was still widely believed. It was Lavoisier who showed how implausible it really was and who correctly identified many of the elements. He demolished one commonly held belief about water quite early in his career. At the time, it was commonly thought that water could be transmuted into earth. After all, watering plants made them grow. It appeared that water was being transformed into a solid substance.

And when distilled water was boiled, a residue was left behind in the vessel used for boiling. Wasn't it obvious that this residue was a kind of earth that had been formed from the water?

It wasn't obvious to Lavoisier, who realized that the idea could be tested by performing an experiment using precise measurements. Lavoisier placed some rainwater, whose weight he had measured, in a sealed glass vessel called a pelican, so-called because it resembles the bird, having two narrow arms that return the substance being distilled to the vessel. Lavoisier weighed the water and the vessel separately and then boiled the water for three months, after which he again weighed the vessel and the water and also the sediment that had formed. He found that the combined weight of the vessel and its contents was exactly the same as it had been originally. The weight of the vessel itself had decreased, however. The weight loss was 12.5 grains more than the weight of the sediment. Lavoisier suspected that the 12.5 grains consisted of material that was now dissolved in the water. If the water contained some of the dissolved vessel, its density (the weight of a cubic centimeter of a substance) should be greater than before. So he tested the water and found that the density had, indeed, increased.

There was only one possible conclusion. No water had been changed into earth. The sediment was material that the water had leached from the glass vessel during the prolonged boiling. Lavoisier had not shattered the idea that earth and water were elements, but he had proved conclusively that one was not being changed into the other.

After Lavoisier had developed his theory of combustion, he was able to go a step further. First, with the assistance of the physicist Simon Laplace, he repeated Cavendish's experiment by burning hydrogen and oxygen in a closed vessel. Next, he passed steam over red-hot iron and found that it could be decomposed into hydrogen and oxygen again. Clearly, water was not an element. It was a compound formed from two gaseous elements.

THE BATTLE OVER PHLOGISTON

The scientific revolution that began when Lavoisier announced his new theory of combustion was far from over. Chemists stuck to the phlogiston theory and some of them continued to cling to the old four-element theory as well, objecting that Lavoisier hadn't really shown that water could be decomposed. The hydrogen, they said, could have come from the hot iron over which the steam had been passed.

But Lavoisier continued to investigate the role played by oxygen in chemical reactions and in animal respiration and to elaborate on his ideas. Finally, in 1785 he launched an attack against the old theory. He submitted a memoir entitled "Reflections on Phlogiston" to the Academy. "The time has come," he said, "to explain myself more precisely and categorically on an opinion that I regard as a disastrous error for chemistry." He then went on to outline all the contradictions inherent in the phlogiston theory. Phlogiston, he said, was something that had never even been rigorously defined. Sometimes it caused bodies to gain weight, sometimes to lose it. Sometimes it was free fire; sometimes it was fire combined with elemental earth. Sometimes it passed through the walls of vessels; sometimes those walls were impermeable to it. Chemistry had to be founded on something better than such a chimerical concept.

Lavoisier's reading of his memoir at the Academy of Sciences was frequently interrupted by violent objections made by listeners hostile to his attempt to discredit the phlogiston theory. According to a Dutch chemist, Martinus van Marum, who was visiting Paris at the time, it was a chaotic session during which Lavoisier and his opponents often tried to make themselves heard at the same time. This, van Marum said, "led to my understanding very little." Lavoisier apparently expected a negative response. In his closing paragraph he said, "I do not imagine that my ideas will be adopted all at once. . . . Time alone will confirm or destroy the opinions I have presented."

Initially Lavoisier was the only anti-phlogistonist in the Academy, but he gradually began to gain converts. The first to accept Lavoisier's ideas were not chemists, however. They were the physicist Laplace and the mathematicians Jacques Cousin and Alexandre Vandermonde. Then in 1787 the chemist Claude Berthollet announced that he had been converted to Lavoisier's theory, and some of the other chemists began to follow.

In 1786 Lavoisier and Guyton de Morveau began a program of reforming chemical nomenclature. They noted that the naming of chemicals was anything but logical. Iron oxide was called "astringent Mars saffron," while sulfuric acid was "oil of vitriol" and zinc oxide was "philosophic wool." Different names were used in different countries, which naturally caused misunderstandings. For example, during a dinner at Lavoisier's house in 1774, Priestley spoke of "red lead," the name given to lead oxide in England. No one understood what he was talking about until one of the other guests suggested that he must mean "minium." Lavoisier, Guyton de Morveau, and their collaborators published the 300-page *Méthode de nomenclature chimique* in 1787. During the course of their work Guyton had become a convert to the new chemistry, and the book was as much a piece of propaganda for the new chemistry as it was a treatise on naming chemicals in a consistent way. For example, zinc calx (which had previously been called "flowers of zinc") became "zinc oxide."

Because the new language incorporated an anti-phlogiston outlook, the book had many detractors. Nevertheless, the nomenclature found rapid acceptance among chemists, who realized that it was far superior to the chaotic assemblage of names used before. Today, Lavoisier's and Guyton's system of nomenclature is still the international language of chemistry.

In 1789 Lavoisier and his colleagues began publishing a new journal called *Annales de chimie*. They founded it, not as a vehicle for anti-phlogiston propaganda, but because they wanted a journal in which papers could be published quickly. Scientific writing submitted to the Academy often did not appear in print for two years. Naturally, contributors to the journal tended to be adherents of Lavoisier's theories, so the publication of the journal had the effect of spreading doubt about phlogiston. The journal was distributed internationally and was quite successful, especially in France and England. Lavoisier's ideas and the new chemical nomenclature quickly became known throughout Europe.

Lavoisier's last salvo was an exposition of the new chemistry titled *Traité élémentaire de chimie (Elements of Chemistry)*. He spent most of 1788 writing it, and it appeared in 1789. In this book, Lavoisier listed some 33 substances he believed to be elements. They were divided into four groups according to their properties. Light, "caloric" (heat), oxygen, nitrogen, and hydrogen were placed in the first group (Lavoisier believed that light and heat were substances), six non-metals were placed in the second, 17 metals in the third, and five "earths" in the last. For the most part, Lavoisier turned out to be right. He listed almost all the elements then known and included only three that are now known to be compounds. The idea that all substances were composed of earth, air, fire, and water was rapidly being forgotten.

JEAN-PAUL MARAT

Jean-Paul Marat was born in a small town in Switzerland in 1743, the same year as Lavoisier. After completing his education in 1760, he spent the next two years as a tutor in Bordeaux. Then, after deciding to pursue a career in medicine, he spent three years in Paris studying. Marat went to London in 1765, and in 1775 he obtained a medical degree from the University of Saint Andrews in Scotland, an eighteenth-century diploma mill, without ever visiting that institution. He returned to Paris in 1777 and quickly acquired a wealthy clientele after successfully treating the Marquise de Laubspine for a lung infection.

In 1778 Marat began a series of experiments on fire, electricity, and light that he hoped would gain him admission to the Academy of Sciences. In 1778 he wrote a paper on his experiments and submitted it to the Academy. The two members assigned to evaluate Marat's work, G. B. Sage and Jean-Baptiste Le Roy, reported favorably on the experimental methods he had devised, though they took no stand on his theoretical conclusions. This encouraged Marat to submit a second paper in 1779. This time the response was more negative. The members who evaluated the paper reported that they were not convinced that Marat's experimental results justified his conclusions. Furthermore, these conclusions contradicted well-established optical laws. They recommended that the paper not be accepted.

Le Roy, who had been a referee of both papers, told Marat that, given the topics he was investigating, he would have to convince Lavoisier, who was by now a leading member of the Academy, that his experiments were significant. Then in 1780 Lavoisier informed his colleagues in the Academy that Marat was claiming that the Academy supported some of the conclusions in the second paper. Because this was not true, the Academy publicly denied the claim. Marat had no trouble discovering who had denounced him. Lavoisier had made himself an enemy.

Marat began a campaign of rhetoric against the Academy, and against Lavoisier in particular. In a pamphlet titled *Modern Charlatans, or Letters on Academic Charlatanism,* published in 1791 but written earlier, Marat wrote:

> At the head of them all would have to come Lavoisier, the reputed father of all the discoveries that have made such a splash. Because he has no ideas of his own, he makes do with those of others. But since he almost never knows how to appreciate them, he abandons them as rashly as he took them up, and he changes systems as often as he changes his shoes. In the space of six months, I saw him adhere, one after the other, to the new theories of matter of fire, ignited fluid and latent heat. In a still shorter space of time, I saw him develop a passion for pure phlogiston and then ruthlessly proscribe it. A while ago, after Cavendish, he found the precious secret of making water with water. Next, having dreamed that this

liquid was only a mixture of pure air and inflammable air, he transformed it into the king of fuels. If you asked me what he did to be so much lauded, I would answer that he managed to get himself an income of 100,000 livres, that he produced the plan to turn Paris into a vast prison and that he changed the terms acid into oxygen, phlogiston into nitrogen, marine into muriatic and nitrous into nitric and nitrac. These are his claims to immortality. Trusting in his great deeds he is now resting on his laurels.

PUBLIC SERVICE

In 1775 Lavoisier was appointed to a government post. He was made one of the four commissioners of the Gunpowder and Saltpeter Administration, which was responsible for the production of gunpowder and acquisition of one of its essential ingredients, saltpeter (potassium nitrate). At the time, the supplies of saltpeter in France were inadequate and the French were forced to buy it from the Dutch, who imported it from India.

Immediately, Lavoisier set to work finding solutions to the problem. He knew a lot about mineralogy so he was able to find new sources of saltpeter in limestone formations, and he improved the techniques of extraction. At the same time he performed chemical experiments in order to better understand how saltpeter was formed. Lavoisier not only increased the production of gunpowder considerably, but he also improved its quality, and French gunpowder soon became the best manufactured in Europe. A year after his appointment, thanks to his efforts, France was producing enough gunpowder for its own needs and had a surplus for export. In 1776 and 1777 France supplied more gunpowder to the rebellious American colonies than it had previously produced in a year. "One can truly say," Lavoisier said with satisfaction in 1789, "that North America owes its independence to French gunpowder."

Lavoisier's next project was one that he inaugurated on his own initiative. In 1778 he bought a large farm at Freschines in the Loire Valley. However, it was not his intention to become a gentleman farmer but to perform studies to see how agricultural output might be increased. At the time, French agricultural production was poor. Yields per acre were much smaller than those in England, for example. Lavoisier wanted to see if he could find ways to increase the French agricultural yield. He found that the output of his newly acquired farm was poor indeed. The soil was good, but the average wheat yield was only about five times the quantity planted as seed. He saw that the property needed more cows and sheep, animals that produced the only kind of fertilizer then known. But if the number of animals was to be increased, there had to be more forage. Creating meadows on which they could graze was no problem, because there was plenty of land lying fallow. But it wasn't clear what would provide the best forage.

Lavoisier immediately began trials. He found that alfalfa tended to be destroyed by a parasite. On the other hand clover and sainfoin (a leguminous herb grown for forage) did well. He bought 20 cows and 500 sheep. This eventually enabled him to triple the amount of manure that was used. He imported breeds of animals he thought would improve his livestock, and he counted and weighed everything that the farm produced so that he would have accurate figures to document the increase in production. Lavoisier grew turnips, rutabagas, beets, potatoes, oats, and wheat. His attempts to raise production met with mixed success. He had good oat harvests, for example, but the production of wheat increased very little because the seed was of poor quality.

After eight years, Lavoisier calculated the rate of return on his investment. It was less than 5 percent. He concluded that this was "no doubt the reason that the well-off farmers living near Paris who manage to save some money prefer to place it in public loans and funds rather than use it for the improvement of agriculture." There was another obstacle to improvement, too, Lavoisier noted: the tax system. When farmers became more affluent and acquired more live-stock, their taxes increased. Because taxes were assessed by the number of livestock owned, any attempt to increase production by using more manure resulted only in an increased tax burden. Lavoisier commented, "All the effect of my incentives [giving animals to his farmers] were destroyed by counter-incentives."

Madame Lavoisier was less enthusiastic than her husband about the agricultural experiment. She much preferred Parisian society to life in the country. After a while she ceased to accompany Lavoisier when he left Paris for the farm. It was during one of Lavoisier's absences, in 1781, that she began an affair with one of the couple's closest friends, the economist Pierre Samuel Dupont de Nemours. Such affairs were common in Paris at the time. Young girls frequently married men much older than they were and after a few years were often attracted to men younger than their husbands. The men, in their turn, frequently took mistresses.

It is not known whether Lavoisier knew about the affair or whether he entered into any liaisons of his own. In any case he and Dupont remained on friendly terms, and they collaborated on work-ing for the improvement of French agriculture through a government body called the Committee on Agriculture, which was created in 1785. While he was on the committee, Lavoisier made numerous proposals for agricultural reform. He spoke of the harmful effects of the tax system that was then in use and criticized other constraints that limited agricultural production. According to Lavoisier, the collective effects of these constraints were disastrous. "Would you believe that a kingdom so fertile," he asked, "which ought to be exporting products of all sorts . . . obtains considerable amounts of them from outside its borders and hence is at the mercy of foreign countries for a large part of the agricultural products for which its soil is best suited?" But in the end, none of Lavoisier's proposals was implemented.

REVOLUTION

When Louis XVI convened the Estates-General on May 1, 1789, he had one thing in mind: getting fiscal reforms that would alleviate a worsening government debt crisis. The Estates, which had not met since 1614, were composed of representatives of the nobility, the clergy, and the Third Estate (commoners). Elections were held between January and April 1789, and 1,200 representatives were elected, 600 for the Third Estate and 300 each for the nobility and clergy. But the representatives of the Third Estate had more in mind than finding money for the Royal Treasury. On June 17 they declared themselves to be a National Assembly and vowed not to adjourn until they had given France a new constitution. The king ordered the Third Estate to disperse. They refused. Then, suddenly, Louis gave in. "Eh bien foutre. Qu'ils restent" ("Oh fuck. Let them stay"), he said, heedless of what might happen next. In July 1785 there were rumors that the king and the nobility were conspiring to overthrow the Third Estate. When Louis called up troops to surround Paris in order to quell any disorders that might arise, the public unrest exploded into revolution. On July 14 a Parisian mob stormed the Bastille, which held only seven prisoners—two of them madmen—but which had long been a symbol of tyranny. By July 15 the revolutionaries controlled Paris. Once again the king gave in; he dismissed the troops surrounding Paris. On August 4 the Assembly abolished aristocratic privileges and tithing to the clergy. There was not yet any desire to depose the king, however. It was the intention of the Assembly to produce a new constitution that would make France into a constitutional monarchy.

When the Revolution began, Lavoisier was still a director of the Saltpeter and Gunpowder Administration and also a director of the Discount Bank, the bank that had propped up the government by repeatedly making large loans to the Royal Treasury. After the Revolution, the bank continued to operate and provided currency by issuing banknotes and continued to lend the government money. On

February 5, Lavoisier took an oath of loyalty and remained in his positions. By this time he was considered a leading financial expert, and the National Assembly appointed him to report on national finances and to suggest reforms in the tax system. The General Farm had been abolished, and something new was needed. Lavoisier set to work, beginning by estimating the gross national product. He found that the gross product was about 2.75 billion livres and the taxable income 1.2 billion. This was the first national accounting ever made.

In March 1791 Lavoisier was named one of six commissioners in charge of the public treasury. He felt uneasy about the idea of holding too many government posts and asked that he be allowed to perform his functions as a commissioner without pay. He had recently been denounced again by Marat, who described him as "son of a land-grabber, apprentice chemist, pupil of the Genevan stock-jobber Necker, a Farmer General, Commissioner for Gunpowder and Salt-peter, secretary to the king, member of the Academy of Sciences, intimate of Vauvilliers,* unfaithful administrator of the Paris Food Commission and the greatest schemer of our times." Lavoisier didn't want to look like he was enriching himself by holding too many paid government positions. It didn't work; Marat denounced him yet again.

WAR

On April 20, 1792, France declared war on Prussia and Austria, which had been persecuting supporters of the French Revolution. Initially the war went badly. The Austro-Prussian army defeated the French. On August 10 a mob, which had been persuaded that the king was in collusion with the invaders, broke into the Tulleries, where Louis was living, and killed several hundred people. On August 13 the king was relegated to a prison. But the rioting did not stop. Between Sep-

---

*Jean François Vauvilliers, a substitute deputy to the Estates-General, who had received secret payments from the king.

tember 2 and September 7, more than a thousand people, who were in prison after being accused of conspiracy with the Austrians and Prussians, were killed. Lavoisier no longer felt safe in Paris, and he left for his farm at Freschines, arriving there on September 15. In August 1789 he had been nearly lynched by a mob, and he had no desire to see something like that happen again.

He remained on the farm for two months before returning to Paris. By this time, the National Assembly had lost control of events, and revolution was again being carried out in the streets. On September 20, a new assembly, the National Convention, met. The king was deposed and a republic was proclaimed. Louis XVI was condemned to death for treason and executed on January 21, 1793.

On June 2, 1793, the radical Montagnards, led by Maximilien Robespierre, Georges Danton, and Marat eliminated the more moderate Girondins from the National Convention and assumed power. On July 13 Marat was assassinated by Charlotte Corday, who was tried and executed on July 17. Marat was embalmed on July 16, and an elaborate funeral, organized by the painter David, was held the following day. As a member of the National Guard, Lavoisier was required to participate in the funeral and pay tribute to the man who had constantly attacked him.

On August 30 the Reign of Terror began. On September 17 the "Law of Suspects" went into effect. It allowed the imprisonment of anyone whose loyalty to the republic or the Revolution was deemed suspect. During the Terror, some 300,000 suspects were arrested. Of these, 17,000 were executed, while numerous others died in prison or were simply killed without trial.

INVESTIGATING THE "CRIMES" OF THE GENERAL FARM

On February 25, 1793, Jean Louis Carra, one of the deputies of the National Convention, proposed that a committee be formed to investigate the "crimes and abuses" of the General Farm and to determine whether all its profits had been legitimate. "There is no time to lose," Carra said. "All those plunderers of public monies, those leeches of the people, those execrable speculators are going to rush to sell their property in France and take flight." However, nothing was done immediately. At the time, the Montagnards and the Girondins were struggling with one another for power. But on September 5, when the Montagnards were securely in control, the personal papers of all of the former farmers were ordered to be placed under seal. On September 28 the order was reversed and the seals were removed. Then, on November 24 the convention changed course again and ordered the arrest of many of the former farmers, including Lavoisier. Nineteen of them were arrested the same day.

At first Lavoisier evaded arrest by going into hiding. Then, on November 28, he gave himself up. He seems to have believed that the farmers could convince a court that all their dealings had been perfectly legal and that, in any case, his record of public service would save him. And of course he was mistaken. Supremely rational minds sometimes fail to appreciate the role played by emotion in human affairs.

TRIAL AND EXECUTION

The farmers were tried on May 8, 1794. They were accused of defrauding the government, enriching themselves at its expense. The indictment listed eight specific charges, all later shown to be false. But the suspect nature of the charges made little difference. Nor did the fact that the court before which the farmers were tried could legally deal only with counterrevolutionary activities, not with the kind of charges that were levied against the farmers.

The trial was concluded in one day. When it ended, the prisoners were sent to the guillotine in the Place de la Révolution. It took 35 minutes to execute 28 farmers. The next day the mathematician Joseph Louis Lagrange commented, "It took them only an instant to cut off that head and a hundred years may not produce another like it."

After the excesses of the Revolution had been played out, it became widely recognized that France had beheaded a great scientist. Thus, after the centenary of the French Revolution, in the 1890s, it was decided that a public statue of Lavoisier should be made. A number of years after the statue was erected it was discovered that the sculptor had used the face, not of Lavoisier but of Jean Antoine Nicolas Caritat Condorcet, who had been secretary of the Academy during Lavoisier's last years. However, the French pragmatically decided that all men in wigs look alike anyway, and the statue remained until it was melted down during World War II. The statue wasn't replaced. However, there is a memorial of sorts to Lavoisier in Paris today. A street in the city's 8th arrondissement is named after him.

CHAPTER 7

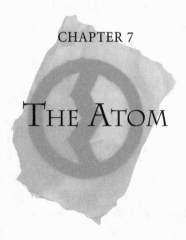

The Atom

T he English chemist John Dalton became one of the most
famous scientists of the eighteenth century. Although he
was known to the public for one idea, that chemical compounds were formed when the atoms of one element joined with the
atoms of another, there was much more than this to Dalton's theory.
He revolutionized chemistry by emphasizing that atoms have relative
weights and that these relative weights can be measured.

The idea that matter is made up of atoms was not new in Dalton's
time. It was known to the ancient Greeks. Even in Robert Boyle's day
the idea that matter was composed of particles, or corpuscles, of some
kind was commonly believed. Dalton revolutionized chemistry, not
by reintroducing this old idea but by formulating a workable theory
of the formation of chemical compounds.

When Dalton propounded his theory, chemistry was not yet a
quantitative science. Chemists did not know how elements combined
or whether they always combined in the same proportions. Indeed,
many chemists believed that the ratio of different elements in a

compound was variable. They knew that metals could be alloyed in whatever proportions one desired and that the quantity of the ingredients of a dye could vary. They did not see why chemical compounds should be any different.

Dalton changed all that by proposing that atoms combine in simple ways. One atom of element A might combine with one of element B. An atom of element C might link up with two or three of D. The number of atoms in a molecule, Dalton said, was always a small whole number. The proportions could not vary; one couldn't have 1.3 or 1.4 atoms of one element combining with 2.4 atoms of another. Furthermore, Dalton said, the atomic theory allowed one to determine the relative weights of the atoms. For example, if nitrogen and oxygen combined to form the oxide NO, and the nitrogen and oxygen were weighed, then one could determine the relative weights of the nitrogen and oxygen atoms by weighing the quantities of nitrogen and oxygen that combined with one another. It might be impossible to determine how much individual atoms weighed, but one could establish, for example, that the atoms of one element weigh three-quarters or seven-eighths as much as those of another. This made it possible to determine the exact chemical composition of chemical compounds.

A QUAKER SCHOOLTEACHER

John Dalton was born into a Quaker family on September 5 or 6, 1766, at Eaglesfield in Cumberland. Because his birth was never registered, there is some uncertainty about the exact date. John's father Joseph eked out a living as a weaver. In those days weaving was a cottage industry and his earnings were meager, so to bring in a little extra money Joseph's wife, Deborah, sold paper, ink, and quills from the family's two-roomed cottage. One room in the cottage contained Joseph's loom; it doubled as a workshop. The other was used for sleeping.

If Dalton's family had not been Quaker, he might have become a weaver himself, or lived out his life as a farm laborer, or followed some other menial occupation. However, the Quakers were great believers in education, and they founded numerous schools for Quaker children. The Quakers considered each member of their sect, woman or man, to be a priest. Thus their schools were coeducational, which was uncommon at the time. Unlike other English schools, the Quaker schools included science in their curricula in the belief that this would help students to better understand God's plans for the natural world.

As a child Dalton was sent, along with his older brother Jonathan, to Pardshaw Hall School, about two miles from his home. There he had an excellent teacher in John Fletcher, who did not believe that students should be force-fed Latin, as they were in most English schools, and who encouraged Dalton to pursue his interest in mathematics. Dalton pursued mathematics tenaciously. If he encountered a problem he couldn't solve at first, he continued to work at it until he found an answer.

Dalton's father was apparently not a hard worker; one of his contemporaries described him as "feckless." In any case, weaving was not an especially remunerative profession and Dalton was forced to leave school at the age of 11. The time had come for him to earn his own keep. At first he worked on a relative's farm. Then, when Fletcher left Pardshaw Hall, Dalton opened a school of his own at the age of 12. Despite his youth, he quickly found himself with a fair number of pupils. Teachers were few in rural areas, and he was offering the only education then available in Eaglesfield.

Dalton held classes in an old barn at first. Later the school was moved to the Dalton cottage and then to the Quaker meetinghouse. His earnings as a schoolmaster were not great, and he supplemented them by continuing his farm labor in his spare time. Some of his pupils were older than he was, and he sometimes had trouble maintaining discipline. The pupils sometimes even threatened to fight him. According to one story, Dalton once tried locking his students in until they finished their work, and they broke the windows in order to escape. But, like many stories about Dalton, this one should be taken with at least a small grain of salt. Dalton's first biographer was not very interested in his subject, and he wrote a hastily produced book containing anecdotes that were based on nothing but hearsay. Many of these anecdotes were repeated by later biographers who made no attempt to verify their accuracy.

Dalton gave up on the school after two years and returned to farm work. At the same time he continued with the self-education that had begun when he left school. His studies were furthered by a wealthy Quaker, Elihu Robinson, who had become interested in Dalton while he was still a struggling schoolmaster. Robinson, who had a great interest in natural philosophy, allowed Dalton to use his extensive library, and tutored him and another boy in mathematics.

In 1781, when Dalton was 15, he joined his older brother Jonathan as an assistant at a Quaker boarding school in Kendal, 45 miles away, which was headed by their first cousin, George Bewley. When Bewley retired in 1785, the two brothers bought his school, getting a mortgage of £150 to do so. Dalton spent the next 12 years teaching there. The income he got from teaching during these years was, naturally, greater than what he had earned while running the school in Eaglesfield. However, it was still quite modest, and he and his brother supplemented it by performing such tasks as collecting rents for landlords, drawing up wills, and performing whatever odd jobs came their way.

It was while he was in Kendal that Dalton made the acquaintance of the wealthy blind scholar John Gough. Though he had been brought up a Quaker, Gough had since become a Unitarian. He had been blinded by smallpox at the age of two, and he suffered from epilepsy as well. Nevertheless, he learned the classics, Latin, Greek, and French, and had a wide-ranging knowledge of natural philosophy. He was especially interested in botany and was said to know every plant in the district by touch, taste, and smell. He could solve mathematical problems in his head, and he published papers on both mathematics and natural philosophy. William Wordsworth, who knew Gough, described him in the following manner in his poem, *The Excursion*:

> Methinks I see him now, his eyeballs roll'd
> Beneath his ample brow—in darkness pained,
> But each instinct with spirit, and the frame
> Of the whole countenance alive with thought,
> Fancy, and understanding, whilst the voice
> Discoursed of natural or moral truth
> With eloquence and such authentic power,
> That in his presence humbler knowledge stood
> Abashed, and tender pity overawed.

THE PROVINCIAL SCIENTIST

Gough adopted Dalton as his protégé, tutoring him in Latin, Greek, and natural philosophy. He gave Dalton access to his library and to his collection of scientific instruments. In return, Dalton became Gough's amanuensis and reader and prepared diagrams for him. Dalton later paid tribute to Gough in the preface of one of his books, writing, "To one person, more particularly, I am peculiarly indebted."

Encouraged by Gough, Dalton began keeping a meteorological journal and took an interest in botany. He kept records of barometric readings, temperatures, rainfall, and wind speeds. He collected specimens of plants and dried and pressed them. Meanwhile, he observed the metamorphoses of caterpillars and observed the effects of placing maggots, mites, and snails in a vacuum or immersing them in water. Dalton pursued his scientific work with enthusiasm, but he made no significant discoveries. It looked as though he was developing into a competent but unremarkable provincial scientist.

As he approached the age of 30, Dalton began to feel dissatisfied with his lot in life. He would have liked to marry and raise a family, but his income as a schoolmaster was not enough to support one. Even as a bachelor, he constantly had to find ways to earn a little extra money to support himself. He began to wonder if he might not train for some more lucrative profession. Two that came to mind were law and medicine. He thought that the difficulties—mostly financial— associated with studying either were not insurmountable.

Feeling in need of advice, he wrote to his friend Elihu Robinson and to his uncle Thomas Greenup, who was a London barrister. Robinson's reply was discouraging. Dalton would do better to remain a schoolmaster, Robinson said; he seemed well suited for that role. Greenup wrote in a different vein. "As to the two professions of law and physic," he said, ". . . I think they are totally out of the reach of a person in thy circumstances." Greenup went on to say that if his nephew was tired of being a schoolteacher, he might set his sights on some lesser profession, such as that of an apothecary. Dalton went on with his teaching and scientific studies, and after a while he gave up the idea of changing professions. By this time he had learned enough about meteorology to write a series of essays on the subject, which were published as a book, *Meteorological Observations and Essays*, in 1793, after he had left Kendal. It was a competent work, but contained little that was new. In Kendal, Dalton really didn't have access to much of the scientific literature on the subject. He did make one contribution however, suggesting that the aurora borealis was an electrical and magnetic phenomenon. As Dalton was later to discover, he was not the first to have this idea. However, his scientific arguments were more formidable than those advanced by earlier writers.

MANCHESTER

In 1792 Gough recommended Dalton for a teaching post at New College, a dissenting academy in Manchester. The college offered him a post, beginning in 1793, and Dalton accepted it without hesitation. In his new position he was expected to teach mathematics, physics, bookkeeping, and chemistry. The following year he was elected to membership in the Manchester Literary and Philosophical Society, which he continued for 50 years, until his death in 1844. During that time, he read more than 100 papers, of which 52 were printed.

The first paper he read was titled "Extraordinary Facts Relating to the Vision of Colours: with Observations by Mr. John Dalton," and it was the first scientific account of color blindness. Dalton suffered from protanopia, sometimes called "Daltonism," the most common kind of color blindness. He was unable to see colors at the red end of the spectrum. Until he was about 26, he was unaware that his vision differed from that of people with normal eyesight. However, when he began to study botany in 1790, he had to pay more attention to colors than before. He found that to his eyes, purple, pink, and crimson all seemed similar to blue, and he often had to ask others whether a flower was blue or pink. In 1792 he happened to look at a pink geranium by candlelight. In daylight this flower appeared to him to be sky blue. But when illuminated by a candle, it seemed not to be blue at all. The color was more like what he was accustomed to call red. So he asked some of his friends whether they perceived any changes in the color of the flower when it was illuminated by a candle. They didn't. He then made up a set of colored ribbons and worked up some tests for color matching and distinguishing among colors. He sought out others with the same type of vision defect, often his pupils. During one term he found that two of his 25 students were color-blind and during another term, one out of 25.

Dalton was unable to correctly explain the cause of protanopia. At the time, the mechanisms of color vision were not yet known; nothing was known of the light-sensitive pigments in the retina. So he hypothesized that the fluid inside his eyes was colored blue and that it absorbed red rays of light. In order to test this theory, he specified in his will that an autopsy be performed and his eyes examined after his death. Of course, no such blue coloring was found. It wasn't until a century and a half after his death that DNA samples could be taken from his eyes, which had been preserved. It was then found that Dalton lacked the genes for producing retinal pigments that were sensitive to red light.

Dalton never married, saying during the latter part of his life that he had never had time for marriage. However, he did experience strong attractions to women from time to time. He described one of his infatuations in a long letter written to Elihu Robinson in 1794. The following is an excerpt:

> The occasion was this: being desired to call upon a widow, a Friend, who thought of entering her son at the academy, I went, and was struck with the sight of the most perfect figure that ever human eye beheld, in a plain but neat dress; her person, her features, were engaging beyond all description. Upon inquiry after I found that she was universally allowed to be the handsomest woman in Manchester. Being invited by her to tea a few days after, along with a worthy man here, a public Friend, I should have in any other circumstances been highly pleased with an elegant tea equipage, American apples of the most delicious flavor, but in the present these were only secondary objects. Deeming myself, however, full proof against *mere beauty*, and knowing that its concomitants are often ignorance and vanity, I was under much apprehension; but when she began to descant upon the excellence of an exact acquaintance with English grammar and the art of letter-writing; to compare the merits of Johnson's and Sheridan's dictionaries; to converse upon the use of dephlogisticated marine air in bleaching; upon the effects of opium on the animal system, &c., &c., I was no longer able to hold out, but surrendered at discretion. During my *captivity*, which lasted about a week, I lost my appetite, and had other symptoms of *bondage* about me, as incoherent discourse, &c., but have now happily regained my freedom.

None of Dalton's attractions ever led to anything serious. His salary at New College was not generous, and he didn't feel that he was financially able to marry until he was past the middle of his long life. By that time, years of bachelorhood might have caused him to become set in his ways. But of course not having a family did have certain advantages. It allowed him to devote long hours to scientific work.

THE DECLINE OF NEW COLLEGE

Near the end of the eighteenth century, enrollment at the dissenting academies began to decline, mainly because the number of dissenters was also declining. In May of 1800 the trustees of New College voted to "curtail for the present the scope of the institution" and to attempt to relocate it elsewhere. Dalton resigned his position the following month. He had no desire to leave Manchester, and he realized that he was now well enough known in the community that he could support himself by tutoring private pupils.

The Literary and Philosophical Society offered him the use of a downstairs room for teaching and experimentation, and Dalton accepted. Soon he was giving lessons six days a week in mathematics, calligraphy, grammar, chemistry, and physics. He supplemented the income from this by lecturing and doing chemical analyses for local manufacturers. He earned enough to live in reasonable comfort and to buy the scientific equipment he needed.

During this period, Dalton did research in meteorology, physics, and chemistry, placing special emphasis on experiments concerning gases, the atmosphere, and quantities of water vapor in the air. Though much of the work that he did during this period was note-worthy, it was far less significant than the atomic theory that was gradually taking shape in his mind.

THE ATOMIC THEORY

In 1803 Dalton read a paper to the Manchester Literary and Philosophical Society in which he mentioned his atomic theory for the first time. But his reference to it was cryptic. The paper dealt with the solubility of different gases in water. After conjecturing that the solubility might depend on the size of the particles of which the gases were composed, Dalton went on:

An enquiry into the relative weights of the ultimate particles
of bodies is a subject, as far as I know, entirely new: I have
lately been prosecuting this enquiry with remarkable success.
The principle cannot be entered upon in this paper; but I
shall just subjoin the results, as they appear to be ascertained
by my experiments.

This was one of the oddest announcements of a discovery in the
history of science. Dalton announced his discovery of his atomic
theory while declining to say what the theory was. Five years were to
pass before he rectified this omission. The first printed account of his
theory was given in his book, *A New System of Chemical Philosophy*,
which was published in 1808. Only 5 of its 220 pages were devoted to
an exposition of his theory. However, the atomic nature of matter
was assumed throughout the book.

Dalton theorized that all chemical elements are made up of small
particles called atoms. He assumed also that all the atoms of any given
elements are exactly alike but different from the atoms of other
elements. Finally, chemical combination happens when one or more
atoms of one element are joined to one or more atoms of another.
For example, one atom of element A might combine with one of
element B. Or one atom of element C might link up with two of
element D. According to Dalton, other small-number combinations
are also possible. For example, in some cases two atoms of one element might link up with three atoms of another. He also stated that,
unless there was some evidence to the contrary, one should assume
that atoms combine in the simplest possible manner. This led quite
naturally to the conclusion that water is made up of molecules containing one atom of hydrogen and one of oxygen. In other words, its
chemical formula is HO.* Of course this is incorrect. However, given

---

*Dalton didn't actually write out formulas like "HO." At this time, chemi-
cal notion had not yet been standardized, and Dalton used pictorial symbols
of different kinds to represent the atoms of different elements.

the state of chemical knowledge at the time, it was a perfectly reasonable conclusion. In any case, Dalton didn't insist that this was the only possibility. In 1810 he stated that the chemical formula for water might also be either HO_2 or H_2O.

At the time there was no way of determining how big atoms were or how much individual atoms weighed. However, as I pointed out at the beginning of this chapter, Dalton's theory did make it possible to determine the relative weights of atoms. For example, there are two oxides of carbon, called carbon monoxide (CO) and carbon dioxide (CO_2) today. If you weighed the quantities of carbon and oxygen that produced these compounds, you could determine the relative weights of carbon and oxygen atoms. When Dalton performed this calculation, he (correctly) found that a carbon atom weighed three-quarters as much as an oxygen atom. However, Dalton's assumption that water molecules had the formula HO caused him to calculate incorrect figures for the weight of hydrogen atoms relative to other elements.

Perhaps it is worth emphasizing again that the importance of Dalton's theory didn't lie in the assumption that matter is composed of indestructible atoms. That was a very old idea. On the contrary, his idea was significant because it was a theory that explained how chemical compounds are formed and because the idea of atoms with different relative weights made it possible to turn chemistry into a quantitative science. As long as chemists held on to the old idea that elements could combine with one another in a variety of different proportions, they could describe chemical reactions only in a qualitative manner. It was Dalton who changed all this.

REACTIONS TO THE THEORY

Dalton's theory hardly won immediate acceptance, and debate about it continued for decades. Chemists generally accepted the idea that elements combine in fixed proportions, and they found the idea of relative weights extremely useful. However, many doubted the reality of Dalton's atoms. Atoms, after all, were too small to be seen, and there was no evidence for their existence. Chemists in England were somewhat more receptive than those on the continent to the idea of tiny invisible atoms. However, even in England, Dalton's theory evoked formidable opposition. The prominent English chemist, Humphrey Davy, was just one of many who were skeptical of Dalton's ideas.

However, the theory was too useful to ignore. Even skeptical chemists used it because they realized that it was needed if they were to do quantitative chemistry. Many provisionally accepted the idea that matter behaved *as if* it were composed of atoms. Some prominent scientists, such as the German chemist Wilhelm Ostwald and the Austrian physicist and philosopher Ernst Mach, continued to adhere to this view into the twentieth century.

It was only in 1905 that the reality of atoms was finally demonstrated. In that year, the same year that he published the first papers on his special theory of relativity, Albert Einstein published a paper on Brownian movement, the irregular motion of small particles suspended in a liquid. Einstein showed that the patterns of movement that were observed could be explained only by assuming that the particles are constantly buffeted by the molecules that make up the liquid. Thus, observations of Brownian movement provided evidence that molecules—and consequently atoms—are indeed real.

DALTON'S LATER LIFE

After the publication of his atomic theory, Dalton's fame increased steadily, and he began to receive numerous honors. He was elected to membership in the French Academy of Sciences in 1816, and he became president of the Manchester Literary and Philosophical Society the following year. When George IV announced in 1825 that he would be giving two gold medals annually, the Royal Society, which administered the awards, gave one of them to Dalton. And in 1832 Oxford University awarded Dalton an honorary doctor of civil law degree.

During the latter part of his life Dalton began to engage in a little myth making, presenting himself as a simple, relatively unlettered Quaker scholar. He claimed, for example, that he could carry all the books he had ever read on his back. In reality he had quite a large library for the time. At his death it was found to contain some 800 books. A story about a visitor from France became part of the Dalton myth. A man known only as "Monsieur Pelletan" came to Manchester for the express purpose of meeting Dalton. At the time, most science was still done by wealthy amateurs, and Pelletan undoubtedly expected to find Dalton residing on some grand estate. After making inquiries, Pelletan found Dalton's laboratory, where he encountered Dalton looking over the shoulder of a boy doing arithmetic on a slate. Pelletan asked if he had the honor of speaking to Monsieur Dalton. "Yes," Dalton replied, "Will you sit down while I put this lad right about his arithmetic." A little later, when Pelletan asked to see the laboratory where Dalton had made his great discoveries, Dalton pointed to an area in the corner of the room and said, "Oh, that's all the apparatus I possess."

You should remember, again, that the anecdotes about Dalton might be apocryphal or contain apocryphal elements. However, the deprecating remark about his apparatus, if he really made it, was certainly in character, and it was as misleading as his remark about the number of books he had read. Dalton actually possessed a number of quite expensive scientific instruments. His expenditures on them represented a significant part of his income.

In 1834 Dalton was presented at the court of William IV. Etiquette prescribed that one wear a sword on such occasions, but as a Quaker, Dalton could not do this. The English mathematician Charles Babbage solved the problem by suggesting that Dalton appear in the scarlet robes of an Oxford doctor of law, and Dalton felt quite comfortable about the idea. Normally a Quaker would never wear scarlet but to Dalton's color-blind eyes, scarlet looked just like the Quaker drab he ordinarily wore. And, yes, there is a rather suspect anecdote about this occasion too. When the king asked him, "Well, Dr. Dalton.

How are you getting on in Manchester—all quiet, I suppose?" Dalton is supposed to have replied, "Well, I don't know, just middlin', I think." However, Babbage made no mention of any such conversation in a long letter he wrote about the event, saying only that, "Doctor Dalton having kissed hands, the King asked him several questions, all of which the philosopher duly answered, and then moved on in proper order to join me."

In the same year a group of Manchester residents began a subscription campaign for the purpose of erecting a statue of Dalton. The committee in charge of the project commissioned Francis Chantry, at the time the best-known portrait sculptor in England, to carry out the project. Chantry took twice as long to complete the work as he had promised, but eventually got it done. When the statue was finally delivered in 1838, it was installed in the entrance hall of the Royal Manchester Institution.

DECLINING HEALTH

When the statue was delivered, Dalton was 74 years old and in declining health. He suffered a stroke in 1837 and another, less serious, one the following year. According to his physician, Dalton had become so frail that "to walk across the two intervening streets [from the Literary and Philosophical Society] to his own house in Faulkner Street, leaning on the arm of Mr. Clare [Peter Clare, one of Dalton's friends] was a great exertion." He nevertheless continued to read papers to the Literary and Philosophical Society, and during the years 1839 and 1840 his health improved to the point that he was again able to spend some time performing experiments in his laboratory. In 1841, however, he took a turn for the worse and one of his former pupils had to read a paper to the society for him.

Dalton died on July 17, 1844, after suffering yet another stroke, and Manchester staged an elaborate funeral. According to the chemist Lyon Playfair, "When Dalton died . . . Manchester gave him the honours of a king. His body lay in state and his funeral was like that of a monarch." Dalton's body was placed in a lead coffin inside one of oak, which was then taken to the Town Hall. According to contemporary accounts, some 40,000 people filed past the coffin on the day the room it lay in was open to the public.

The public funeral was held on August 12, when the coffin was taken from Town Hall to a nearby cemetery in a funeral procession that was nearly a mile long. Six horses drew the hearse, followed by the more notable members of the community riding in almost a hundred carriages and the common citizens of Manchester on foot.

In the years following the funeral, Manchester citizens continued to pay tribute to their great natural philosopher. In 1853 a granite monument was erected over Dalton's grave. Scholarships were established in his name at Manchester's new Owens College, and in 1903 the city celebrated his discovery of the atomic theory. It was a fitting tribute to the man who had transformed chemistry.

CHAPTER 8

PROBLEMS WITH ATOMS

J öns Jacob Berzelius, born in Sweden in 1779, was an orphan who was brought up by his stepfather. During his youth he worked on his stepfather's farm, living in a room that was also a storeroom for potatoes. When he grew a little older, he went to the nearby town of Linköping, supporting himself by tutoring. At first he thought of becoming a clergyman, but then became more interested in natural philosophy. Warned by the rector of the school at his graduation that his lack of ambition "justified only doubtful hopes," Berzelius nevertheless went on to study medicine at the University of Uppsala, where he became interested in chemistry and got permission to do experiments in the chemistry professor's laboratory. After Berzelius graduated in 1802, a wealthy mine owner offered him the use of his home laboratory, where in 1803 he discovered a new element, cerium.

In 1807 Berzelius became professor of chemistry and pharmacy at the Carolian Medico-Chirurgical Institute in Stockholm. His duties there were not heavy, and he was able to spend a great deal of time conducting experiments in the institute's laboratory. The papers that he published earned him some renown, and he was elected to the Swedish Academy of Sciences in 1808. It was during that year that Berzelius learned of the work of Dalton. Unfortunately, the wars that were raging in Europe at the time prevented him from getting a copy of Dalton's book, *New System of Chemical Philosophy*, until 1812. However, the secondhand accounts of Dalton's theory that Berzelius read convinced him of the significance of Dalton's accomplishment.

Berzelius's own experiments soon convinced him that Dalton was right to conclude that atoms always combine with one another in small whole-number ratios. Berzelius realized that determining the relative weights of *all* of the elements would be of enormous value to chemistry, because it would then be possible to determine the exact composition of any chemical compound. "Without work of this kind," he said, "no day could follow the morning dawn." Because he knew of no other chemist who was pursuing this line of research, he decided to do it himself.

He had set himself a Herculean task. By this time nearly 50 chemical elements were known. Furthermore, many of the chemicals needed to carry out the work could not be purchased, while many of the others could be obtained only in impure form. Also, much of the chemical apparatus currently available was too crude to be used for the kinds of accurate analyses that he would need to do.

Berzelius purified the chemicals he used, not once but dozens of time. He took lessons from a glassblower so that he could construct his own apparatus. And if he needed an apparatus that didn't exist, he invented it. He was the first to use such now-common laboratory items as filter paper, water baths, desiccators, and rubber tubing. He put these items to good use and spent 10 years studying some 2,000 chemical compounds. Berzelius kept on working methodically, carefully weighing and measuring until he had determined the relative weights of some 45 different elements.

Dalton had arbitrarily assigned a relative weight of 1 to hydrogen, the lightest element. But Berzelius didn't follow the English chemist's example. Instead he used two different systems at different points in his career. Initially he gave oxygen a weight of 100. Then in 1831 he switched to a system in which hydrogen had a weight of $1/2$. Because using Berzelius's actual numbers would be confusing, I will describe some of his results using the modern system, which assigns oxygen a weight of exactly 16, in the account that follows.

Oxygen is by far the most abundant element in the Earth's crust and is a component of numerous common chemical compounds. Thus it was especially important to find the atomic weight of its atoms. Berzelius correctly determined that an oxygen atom was 16 times as heavy as one of hydrogen, and he assigned it a weight of 16.00. This result was not in agreement with that of Dalton, who, assuming that water had the formula HO, had obtained a figure of 7 for oxygen. Berzelius also did not agree with some of Dalton's other determinations. For example, Dalton had concluded that a nitrogen atom had a relative weight of 7, while Berzelius calculated that the figure should be 14.05.

Some of Berzelius's results were remarkably accurate. For example, he obtained a weight of 35.43 for chlorine, which is very close to the modern figure of 35.45. Similarly, he found the weight of lead to be 207.12, which hardly differs from the modern 207.19. However, in some cases, ignorance of the chemical formulas of certain compounds led to values that were far too high. Berzelius assigned atomic weights to the alkali metals (for example, sodium and potassium) and silver that were about twice what they should be. This led, in turn, to incorrect chemical formulas for other compounds. While Berzelius was engaged in this work, he still found time to discover a series of new elements in addition to cerium. These included selenium (1817) and thorium (1828).

Because Berzelius's atomic weight determinations did not always agree with Dalton's, they were greeted with some skepticism at first, especially in England. However, as chemists throughout Europe performed their own atomic-weight experiments, acceptance of his results grew. For the most part the European relative weight determinations confirmed that Berzelius's results were remarkably accurate.

CONFUSION

Berzelius's accomplishment was a great step forward. However, in the mid-nineteenth century, confusion continued to reign. It appeared that the more that chemists knew, the more puzzles that confronted them. First, there was the question of why there were so many different chemical elements. When Dalton died in 1844, about 50 were known. Was the universe really made of 50 different building blocks? The physicists had discovered that fundamental physical laws could be based on simple assumptions. Newton's law of gravitation, for example, could be written using just a few mathematical symbols. Why then should the world of chemistry be so complicated?

Many chemists were still unable to accept the idea of tiny invisible atoms. They realized that something like atomic weights was needed to do analytical chemistry, but they avoided using the term "atomic weight" and spoke of "combining weights" or "proportional numbers" instead. Thus chemical terminology was anything but uniform. Meanwhile, other chemists still did not accept the conclusion of Dalton and Berzelius that elements always combined with one another in certain fixed proportions. The belief that the proportions were variable lingered on to the end of the nineteenth century. Wilhelm Ostwald, who won the Nobel Prize for chemistry in 1909, was probably the most prominent chemist to cling to this idea, but he was not the only one.

As new experiments were performed, the results sometimes seemed to contradict Berzelius's atomic weight determinations. Different chemists began to make different assumptions about how the weights should be calculated. There was confusion also about the terms "atom" and "molecule" themselves. At mid-century these two words were still used interchangeably. Even the discovery of simplicity sometimes proved to be confusing. In 1808 the French chemist and physicist Joseph Louis Gay-Lussac discovered a simple chemical law and chemists did not know how to explain it.

Born in 1778 at Saint-Léonard-de-Noblat in France, Gay-Lussac grew up during the Revolution. His father, a country lawyer, was a liberal monarchist who believed that reforms could be carried out without abolishing the monarchy. These views got him arrested during the Terror, and he was not released from prison until the fall of Robespierre. Though his father's arrest drastically reduced their income, Gay-Lussac's family nevertheless ensured that he got an education. He did so well in school that as he approached the age of 16 he got a government grant that allowed him to attend France's best university, the École Polytechnique. One of the lecturers there was the chemist Claude Berthollet. On a friend's recommendation Berthollet hired Gay-Lussac to work as an assistant in his laboratory, and later invited the young man to work with him in the laboratory that he maintained at his country house. The collaboration continued for a number of years, and when Berthollet retired, Gay-Lussac was appointed to his position at the École Polytechnique.

When he was 25 Gay-Lussac met the German naturalist Alexander von Humboldt. The two soon found themselves discussing the composition of water and decided to perform some experiments. They discovered that when two volumes of hydrogen are chemically combined with one volume of oxygen, exactly two volumes of water vapor are produced. For example, two liters of hydrogen and one liter of oxygen combine to form two liters of vapor. This was not a new discovery; the phenomenon had been noted before. However, Gay-Lussac was inspired to experiment with other gases also. He found that they, too, combine in simple ratios. For example, two volumes of carbon monoxide combine with one of oxygen to produce two of carbon dioxide, while three volumes of hydrogen and one of nitrogen produce two volumes of ammonia gas.

Gay-Lussac announced the results of his experiments in 1808. The reaction was mixed. Although Berzelius accepted the results and used them in his atomic-weight research, Dalton maintained that they couldn't possibly be correct. "The truth is," he said, "that gases do not unite in equal or exact measures in any one instance; when they appear to do so, it is owing to the inaccuracy of our experiments." It isn't surprising that he should have rejected Gay-Lussac's results, because they appeared to be incompatible with his theory and it isn't difficult to see why. Dalton assumed that gases are composed of a large number of individual atoms. If the chemical formula for water was HO, then one volume of hydrogen should combine with one of oxygen to form one volume of water vapor. On the other hand, if the formula for water was H_2O, then two volumes of hydrogen should have combined with one of oxygen to produce one volume of water vapor, not two, as Gay-Lussac had found. Apparently, the only way that the experimental results could be reconciled with Dalton's theory was to conclude that oxygen atoms could split in two and that one-half of an oxygen atom could combine with one atom of hydrogen.

Although chemists wrestled with the problem, they found no logical solution. Further experimentation seemed only to make matters worse. When the French chemist Jean-Baptiste Dumas measured the densities of different kinds of vapors during the 1820s, he achieved results that were even more difficult to reconcile with the atomic theory. The theory couldn't be abandoned; by this time it had become an integral part of chemistry.

A SOLUTION FOUND

Amedeo Avogadro was born in Italy in 1776 into a line of ecclesiastical lawyers. Like his father, Philippe, he trained for a career in the legal profession. Avogadro got a baccalaureate in jurisprudence at the age of 16 and a doctorate in ecclesiastical law at the age of 20. After practicing law for three years, he became interested in natural science and spent years studying chemistry, physics, mathematics, and philosophy. He was soon carrying out chemical experiments with his brother Félice. Their results were significant, and the two brothers attracted some attention from other scientists when they presented a paper to the Academy of Sciences of Turin. In 1809, at the age of 33, Avogadro became professor of physics at the Royal College at Vercelli. In 1811 he published a paper that provided the solution to some of the problems that nevertheless continued to bedevil chemists in 1860. Avogadro's paper had virtually no impact. Not a single scientist commented on it, and most remained ignorant of its publication.

The paper proposed two hypotheses. The first was that any given volume of a gas always contains the same number of molecules provided temperature and pressure are held constant (this is now known as Avogadro's law). The second was that gases such as hydrogen, nitrogen, and oxygen were not large ensembles of individual atoms as Dalton and his successors thought. Gay-Lussac's results, Avogadro believed, could easily be explained if one assumed that those gasses are made up of molecules containing *pairs* of atoms. The chemical formula for oxygen gas, for example, is not O but rather O_2.

Avogadro's discovery caused not the slightest stir. Nevertheless, he continued to perform chemical experiments, mainly in physical chemistry. Then in 1820, when a chair of mathematical physics was created at the University of Turin, the position was given to Avogadro, but he was not to hold it for long. A revolution broke out in Naples and before it was suppressed there was a rebellion in Piedmont also. King Victor Emmanuel I abdicated in favor of his brother Charles Felix, who closed the University of Turin after putting down the uprisings. Avogadro, who was given a very small pension, returned to the practice of law.

When Charles Felix died, his successor Charles Albert, reopened the university, and Avogadro resumed his old post. He continued to teach for 20 years before retiring. When he died in 1856, his groundbreaking ideas were still unknown, even in Italy. His papers, which had been written in French, were not even translated into Italian until the beginning of the twentieth century. However, Avogadro's theory was not to remain in obscurity for long. In 1860 an international conference of chemists was held at Karlsruhe in Germany. Some 140 chemists came to discuss the problems confronting their science. The confusion surrounding atomic weights had still not been cleared up, and there were disagreements about the role played by atoms in chemistry. In the end, the conference participants accomplished very little. Most of the disagreements that existed when the conference began remained when it came to an end. Except for one thing, the attendance at the conference of Stanislao Cannizzaro, the meeting would have been a forgettable event.

Cannizzaro was born in Palermo, Sicily, in 1826. At the age of 15 he began to study medicine at the university in his native city. When he became interested in chemistry, he went to the University of Pisa, and then to Naples. In 1847, when a rebellion against the ruling Bourbons broke out in Sicily, Cannizzaro, who was then 21, rushed home to join the insurrection. When the rebellion failed, Cannizzaro, who had been an artillery officer and a deputy to the newly formed revolutionary Sicilian parliament, was condemned to death. However, he managed to escape to France, where he found work in the laboratory of the Parisian chemist Michel-Eugène Chevreul. He was later able to return to Italy, where he got a post teaching at the National College of Alessandria in northern Italy.

In 1860 Cannizzaro, now teaching at the University of Genoa, heard that rebellion had again broken out in Sicily. He traveled there to participate but this time he arrived too late. The "red shirts" led by Guiseppe Garibaldi had already freed Sicily. It was at about this time that Cannizzarro received an invitation to attend the conference at Karlsruhe. Because there was no revolution to become involved in, he accepted at once.

Cannizzaro was one of the very few chemists who understood the significance of Avogadro's accomplishments, and he had been teaching his students about Avogadro's discovery for years. However, Avogadro was still unknown outside Italy, and Cannizzaro realized that the conference was an opportunity to speak about his theory to an international audience. When he made an impromptu speech about Avogadro's theory, the response was not enthusiastic. It appeared that he did not get his message across. Nevertheless, at the end of the conference, one of the Italian participants distributed copies of a pamphlet Cannizarro had written on Avogadro's ideas and their usefulness for determining correct atomic weights. This time the response was better. For example, one of the younger participants, the German chemist Lothar Meyer, put the pamphlet in his pocket and read it on his way home. "It was as though the scales fell from my eyes," he later wrote. "Doubt vanished and was replaced by a feeling of peaceful clarity." Four years later Meyer discussed Avogadro's ideas in his book, *Modern Theories of Chemistry*. At last the long-neglected work of Avogadro was making its way into the chemical mainstream.

The ultimate acceptance of Avogadro's theory was of enormous significance to chemistry. It did away with the apparent contradiction between Dalton's theory and Gay-Lussac's law, and it cleared up the uncertainties surrounding atomic weights and chemical formulas. "Instead of taking for your unit of atomic weight the weight of an entire molecule of hydrogen," Cannizzaro said, "take rather half of this weight." At the time, hydrogen was usually assigned a weight of 1, so when the confusion surrounding the weight of hydrogen was cleared up, numerous other difficulties rapidly disappeared.

LOOKING FOR PATTERNS

Chemists still didn't know why there were so many different chemical elements or whether any patterns could be found in them. In 1815 the English physician William Prout had proposed the hypothesis that all of the elements were condensed hydrogen. For example, the atomic weight of oxygen was 16. According to Prout, this indicated that 16 volumes of hydrogen had condensed to form this element. Of course Prout's hypothesis was incorrect. However, during the nineteenth century there was no empirical evidence that either supported or contradicted the idea. Consequently, some chemists adopted the theory while many others opposed it.

Many chemists suspected that speculation such as Prout's would lead nowhere. They thought it more fruitful to try to find some order among the chemical elements. It was known that certain elements had similar properties. Potassium behaved very much like sodium, for example. Both were soft, bright, silvery metals, and both reacted with other chemical elements in the same ways. If further clues about the relationships between similar elements could be discovered, perhaps some important insights could be gained.

As early as 1817, the German chemist Johann Döbereiner (who was Goethe's chemistry teacher) had noticed a curious relationship between the oxides of three elements with similar properties. The molecular weight of strontium oxide was exactly midway between those of the oxides of calcium and barium. A decade later Döbereiner noticed a similar relationship among three other related elements. The atomic weight of bromine was also halfway between those of chlorine and iodine. Soon other chemists began to notice simple relationships between the weights of chemically similar elements. This caused the English chemist William Odling to wonder if elements couldn't be arranged in natural groups according to their properties. In 1857 Odling divided the 49 known elements into 13 different groups. For example, he placed fluorine, chlorine, bromine, and iodine in one group. Oxygen, sulfur, selenium, and tellurium made up another, while lithium, sodium, and potassium were placed in a third. Odling's speculations didn't lead to much, however. It was already known that certain elements behaved in similar ways, and there was no obvious reason why the number of different groups should be 13.

In 1862 the French geologist and mineralogist Alexandre Chancourtis arranged the elements on a cylinder that he called a "telluric screw." Placing the elements along a descending spiral according to their atomic weights, Chancourtis found that the elements could be placed in vertical groups so that the property of any given element was similar to the ones above and below it. Chancourtis had discovered a pattern. However, chemists ignored his paper for several reasons. Because Chancourtis was a geologist, he sometimes used geological terms instead of their chemical counterparts. And for some reason, he included some compounds and alloys in his helical array. Finally, an illustration of the telluric screw was omitted from his paper, making it very difficult to understand. Chancourtis did publish the drawing later, but doing so failed to arouse any interest.

John Newlands was a London sugar refiner who had studied at the Royal College of Chemistry and had fought with Garibaldi in Italy. In 1863 and 1864 he published some short papers in which he noted that when the elements were listed in the order of increasing atomic weights, every eighth element had properties analogous to the first element in its group. In order to make his system work, Newlands had to change the order of some of the elements, sometimes placing one that was lighter after one that was heavier. It later turned out that he was justified in doing so; at this time, the atomic weights of the elements had not always been determined correctly.

Newlands called his system a "law of octaves," and he made comparisons between his system and the notes that could be played on a piano. This wasn't conducive to acceptance of his ideas. After all, there was no reason why there should be a relationship between the chemical elements and the notes in the diatonic scale. When Newlands described his scheme at a meeting of the Chemical Society in 1866, the chemist and physicist George Carey Foster sarcastically suggested that Newlands might just as well have arranged the elements alphabetically. Foster's point was that the analogies that Newlands saw could easily be the result of coincidence. Furthermore, Newlands had altered his arrangement more than once and he sometimes assigned 10 elements to an octave rather than eight. In the view of many chemists the analogies Newlands saw were nothing more than the result of his tinkering.

The chemists of the day failed to see that both Chancourtis and Newlands had a fundamental insight: that the key to finding order in tables of the chemical elements was to arrange them in order of increasing atomic weight. Indeed, it wasn't at all obvious that this should be the case. It seemed more natural to group elements with similar properties together, without considering atomic weight, as Odling had done.

Chancourtis and Newlands had been groping in the right direction. However, their systems didn't quite work, even though they gave some tantalizing hints about order in the chemical world. Their methods were too rigid, and they didn't allow for the inclusion of as-yet-undiscovered elements. Both men had caught glimpses of a periodicity in the chemical elements, but they weren't able to work out important detailsw or persuade their contemporaries of the significance of what they were trying to do.

CHAPTER 9

THE PERIODIC LAW

I n 1884 the Scottish chemist Sir William Ramsay went to London to attend a dinner honoring William Perkin, the discoverer of mauve, the first synthetic dye. Arriving early, he encountered "a peculiar foreigner, every hair of whose head acted in independence of every other." When the foreigner approached, bowing, Ramsay said, "We are to have a good attendance, I think?" Discovering that the man didn't speak English, Ramsey asked him if he spoke German. "Ja, ein wenig" ("Yes, a little"), the foreigner replied. "Ich bin Mendeleev."

Ramsay related later that, "He is a nice sort of fellow, but his German is not perfect. He said he was raised in East Siberia and knew no Russian until he was seventeen years old. I suppose he is a Kalmuck or one of those outlandish creatures." Dimitri Mendeleev* wasn't a Kalmuck (the Kalmucks were Buddhist Mongols), but he did have

---

*Like many other Russian names, "Mendeleev" can be transliterated in several different ways. "Mendeleyev" has been the most common, but "Mendeleev" is a bit more accurate.

something of an outlandish appearance. He dressed reasonably well, but his unkempt white hair fell to his shoulders. He was in the habit of having his hair and beard cut once a year, and to some he might have looked more like a Siberian shaman than a distinguished chemist.

It was Mendeleev who discovered the periodic law, a principle that describes the periodicities that are observed in the properties of the chemical elements. This enabled him to predict the existence of as-yet-undiscovered elements, to predict their atomic weights, and to describe their chemical and physical properties as well. It was Mendeleev who found the natural order in the elements that his predecessors Newlands, Chancourtis, and Odling had been seeking. Mendeleev couldn't explain why there were so many elements; he didn't even try to do that. But he did discover the existence of striking patterns.

SIBERIA

Mendeleev was born in Tobolsk in western Siberia in 1834. His father was director of the local high school and the family lived comfortably. However, all that suddenly changed when Mendeleev was still a child. His father Ivan became blind from cataracts and had to resign his post. He went to Moscow to have them removed and had an operation that was a partial success. He could see again and recognize faces. However, he was still unable to read books and thus couldn't resume his career, and he did not live long after the operation. After his death, of tuberculosis, his widow got his pension, but the money was far from enough to support the family.

Fortunately, Mendeleev's mother, Maria, was a remarkable woman. Her father had established the first printing press and published the first newspaper in Siberia. Maria apparently inherited some of his determination. When her brother wanted to reopen a rundown glass factory that he owned at Aremziansk, a village 17 miles from Tobolsk, he wrote to ask her to recommend a capable manager and she recommended herself. Her brother gave her the job, and the family moved to Aremziansk.

One of the people living in Aremziansk was Nicole Bassargin, who had been a Decembrist. The Decembrists were a society of revolutionaries that was organized in the 1820s and made up primarily of members of the upper class with military backgrounds. Their goals were the institution of a constitutional monarchy and the freeing of the serfs. On December 14, 1825, they staged an uprising in St. Petersburg, which was suppressed by the new Tsar, Nicholas I. Five of the leaders were executed, and more than a hundred were sentenced to hard labor in Siberia.

After making the acquaintance of the Mendeleev family, Bassargin courted and later married Mendeleev's older sister Olga. Since his release after serving his sentence he had built up a large library, and he had a great interest in science. During his visits to the family, he often discussed science with Mendeleev. After his marriage, he settled in the nearby village of Yaloutorvsk. But his visits to the Mendeleev home continued, and he continued to tutor Mendeleev in the sciences.

In 1849, the same year that Dostoevski was arrested and sentenced to a Siberian labor camp, disaster struck the family again. The glass factory caught fire one winter night and burned to the ground. But the 57-year-old Maria Mendeleev again showed her determination. She decided that she would take her son to Moscow in order to get him admitted to the university there. She was convinced that he had great potential.

At the time, Russia was backward compared to the nations of Western Europe. However, it had one of the best educational systems on the continent. It boasted good universities and a government scholarship program designed to aid promising youths who might otherwise not get university educations. Maria intended to take advantage of this to secure an education for her son. He would, of course, have to pass an examination to get a scholarship. But he was a bright young man, and she had no doubt that he would pass.

It wasn't an easy journey. The Trans-Siberian Railway did not yet exist. Fifteen years previously a large network of railroads had been proposed, but Russia's minister of finance, who believed that rail travel was a threat to "public morals," rejected the idea. Railroads, he said, "encouraged frequent purposeless travel, thus fostering the restless spirit of our age."

ST. PETERSBURG

In the absence of railroads, mother and son hitchhiked the 1,300 miles to Moscow. When they arrived, it appeared at first that their journey had been in vain. The University of Moscow had a quota system for admitting students from the provinces, and Siberia had not yet been given a quota. Thus Mendeleev could not take the examinations for a scholarship, or even apply to enter the university. But Maria Mendeleev didn't even consider returning to Siberia. Instead, she set off with her son, and a daughter, on the 400-mile journey to St. Petersburg. Shortly after arriving she discovered that the director of the St. Petersburg Pedagogical Institute, a college that trained teachers, was Ivan Pletnov, an old friend of her late husband. Pletnov allowed Mendeleev to apply for admission to the Institute and to take the scholarship examinations, which he passed. As a scholarship student, Mendeleev lived at the Institute, which provided room, board, books, and school uniforms. When he graduated he would be required to teach for eight years. However this did not seem an onerous requirement.

Mendeleev studied mathematics, physics, and chemistry at the Institute. Shortly after he was admitted, his mother died. A year later his sister Liza, who had accompanied them to St. Petersburg, died also. But Mendeleev continued to work hard at his studies, graduated first in his class in 1855, and was awarded a gold medal. However, as was the custom, the examiners praised not Mendeleev but his chemistry professor A. A. Voskresenski, for having produced such a fine pupil.

At this time, Mendeleev was not in good health. He was coughing blood, and the physician at the Institute hospital diagnosed tuberculosis, giving Mendeleev six months to live. The doctor thought, however, that a warmer climate might delay the progress of the disease, so the Institute authorities arranged a teaching post for Mendeleev at Simferopol in the Crimea. He arrived in Simferopol only to discover that he didn't have a job after all. The Crimean War was raging. Simferopol had been turned into a kind of military camp, and the school at which he was to have taught was closed. However, the town officials gave him a month's salary.

There was no reason for Mendeleev to stay in Simferopol, so he set off for Odessa, a large city in the Ukraine, located on the Black Sea. There he consulted Dr. Nicolai Pirogov, who had formerly been a physician at the Pedagogical Institute. Pirogov discovered that Mendeleev didn't have tuberculosis at all. According to Pirogov, the coughing of blood had been caused by a "valvular defect" in Mendeleev's heart. Mendeleev might cough blood again in the future, Pirogov said, but there was no reason why he shouldn't live well into old age.

Mendeleev soon got a post teaching mathematics and physics at an Odessa high school. Granted permission to use the library and laboratory at Novorossisk University, he launched into research for his master's thesis. By May of 1856 he had completed a first draft. He revised it during the summer and was awarded a master of physics and chemistry degree by the University of St. Petersburg in September.

STUDIES IN EUROPE

Six weeks later the 22-year-old Mendeleev won an appointment at the university as a *privat-docent*, a kind of unsalaried assistant professor who received part of the fees paid by his students. During the mid-nineteenth century, Russia was somewhat backward scientifically, so Mendeleev had no contact with leading chemists. However, in 1859 he got a government grant to pursue further studies in France and Germany. On the advice of his friend, the chemist and composer Alexsandr Borodin, he went first to Paris, where he studied with Henri Regnault, the noted French chemist and physicist. Then he went to Heidelberg where he worked in the laboratory of Bunsen and Kirchhoff.

Kirchhoff and Bunsen's laboratory was an important center of scientific activity. This should have made it an ideal place for a young scientist to work. However, Mendeleev did not get along well with Bunsen, who supervised his experimental work. Mendeleev wanted to pursue some ideas of his own, while Bunsen expected him to perform the assignments that were given him. One day Mendeleev's frustration led to an outburst of anger, and he abruptly walked out. Though he continued to attend Kirchhoff's and Bunsen's lectures, he made up his mind to set up his own laboratory in his lodgings.

The University of Heidelberg, which had been founded in 1386, was a famous institution, and it had quite a few Russian students. It attracted Russian visitors as well. One of these was the novelist Ivan Turgenev, who might have had Mendeleev in mind when he later wrote of "a young Russian chemist living in Heidelberg who was praised by all who knew him as an uncommon talent." One of the other people Mendeleev encountered in Heidelberg was his friend Borodin, who had come to Western Europe for further study after receiving his doctorate in chemistry in 1859. Mendeleev and Borodin took short trips to Italy together during university holidays, but then Mendeleev began to see less and less of his friend. Borodin had become enamored of a young Russian girl who was in Heidelberg for her health. The two were married on their return to Russia.

In December 1860 Mendeleev attended the congress of chemists in Karlsruhe, where he heard Cannizzaro speak and read his pamphlet. The congress took place near the end of Mendeleev's stay in Europe. Some three months later he was summoned to St. Petersburg to teach a course in organic chemistry. He was now 27 years old, and his publications in German and French scientific journals were making him known outside of Russia. However, he had not yet embarked on the research that was to make his name known throughout the world.

RETURN TO ST. PETERSBURG

News traveled slowly in those days, and it wasn't until he returned to Russia that Mendeleev learned that the new Russian tsar, Alexander II, had just freed the serfs. Alexander, who had come to the throne in 1855, had been appalled by Russian defeats in the Crimean War. Seeing them as a sign of Russia's backwardness, he instituted a series of reforms designed to make Russia more like the nations of Western Europe. He vastly expanded Russia's railroad system, reformed the judicial system, set up local assemblies and village schools, reorganized the military, freed the serfs, and distributed land to the freed peasants. Alexander was no democrat, however, and had every intention of continuing to be an autocratic ruler. In 1862 he reacted to the spread of radical doctrines by instituting repressive police measures. After an assassination attempt in 1866, the role of the secret police was increased even more. Thus the Russia that Mendeleev returned to combined liberal economic reforms with increasing political repression.

Mendeleev was not very concerned with political matters at this time, however. He was far too occupied with his scientific work. He wrote a doctoral dissertation, completed a 500-page textbook on organic chemistry, and in 1862 he married. Mendeleev's older sister Olga arranged the marriage, which Mendeleev initially agreed to. However, once he began to know his future wife a little, he began to have doubts. When he conveyed his doubts to his sister, she became angry. In those days engagements were regarded very seriously, and if a man broke an engagement, he cast dishonor upon himself while doing a great wrong to his betrothed. Olga let Mendeleev know what she thought in no uncertain terms, and he went ahead with the marriage.

The marriage was not a happy one. The couple argued constantly. Mendeleev's submersion in his work might also have contributed to the conflicts with his wife. In addition to lecturing at the university on organic chemistry, he taught chemistry at a military school and continued with his own research. Realizing that Russian scientists were unaware of the advances in chemistry that had recently been made in Europe, he translated German works on the subject. He became interested in scientific agriculture and analyzed soil samples for the Imperial Free Economics Society. In addition to all this, he somehow found the time to travel to Baku on the Caspian Sea to study the methods of producing oil in order to make recommendations to the owner of a refining company. This was no light undertaking. In those days of no railroads and poor roads, it was an arduous journey, yet Mendeleev readily accepted the assignment.

He got his doctorate in 1865 and shortly thereafter was made a full professor at the University of St. Petersburg. He was now 31 years old. One of his students was Prince Peter Kropotkin, who was later to become a prominent anarchist leader. But Mendeleev had many other students too; as many as 200 would come to see him lecture. His popularity might have been because he taught in an informal manner, often relating anecdotes and digressing into such topics as astronomy, meteorology, geology, biology, and agronomy, even balloon navigation and the use of artillery. The chemical demonstrations that Mendeleev and his assistants prepared certainly made an impression too, especially, one imagines, those that featured explosive reactions.

But of course Mendeleev was much more than a showman. Realizing that no adequate textbook of chemistry existed in Russian, he planned to write his own. At the same time, he continued to ponder a problem that had long concerned him, the fact that chemistry had no central guiding principle. Surely some kind of order existed in the chemical elements, 63 of which were then known. There had to be some pattern.

THE PERIODIC LAW

In 1867 Mendeleev began writing the first of a two-volume text, *Principles of Chemistry*. As he wrote, he found it natural to group elements with similar properties together. In the first volume, he discussed the alkali metals: lithium, sodium, potassium, rubidium, and cesium; the halogens: fluorine, chlorine, bromine, and iodine, and their compounds; and the very common elements: hydrogen, oxygen, nitrogen, and carbon. This raised the question of what elements should be discussed first in the second volume and how they should be grouped together. This caused his thoughts to turn again to the idea of finding an ordering principle that would show how the elements were related to one another.

Mendeleev tackled the problem by making up a card for each of the 63 known elements. On each he wrote the atomic weight of the element and listed its most significant properties. In order to get the most accurate atomic weights available, he wrote to a number of chemists, asking for the figures they had obtained from their measurements. Thus he got information from the Belgian chemist Jean Servais Stas, the French scientist Jean-Baptiste Dumas, the English scientist Sir William Crookes, the Swedish chemist Kruss Nilson, and a professor at Prague University, Bohuslav Brauner.

Mendeleev wrote the weights on his cards as he received them. Then he verified as many as he could with his own experiments. Finally, he arranged the cards in order of atomic weight, beginning with hydrogen, the lightest element and ending with uranium, the heaviest then known. On each card he also noted the chemical properties of the element and certain of its physical properties, such as melting point, density, and malleability. Then he pored over the cards for days, looking for patterns. Finally he pinned the cards on a wall, putting similar elements in horizontal rows. He looked at the table that this formed, made changes, and pinned the cards on the wall again.

Mendeleev arranged the elements into seven groups. Lithium (atomic weight 7) was followed by beryllium (9), boron (11), carbon (12), nitrogen (14), oxygen (16), and fluorine (19). The next element in order of atomic weight was sodium (23), which had properties similar to those of lithium. Therefore, Mendeleev pinned the card for sodium under that for lithium. Six more cards were placed in the second row, ending with chlorine under fluorine. He continued in the same manner until all 63 cards were placed. When he finished, he noticed something very striking, that the properties of these elements "were periodic functions of their atomic weights." In other words, the same kinds of properties were observed after every seven elements.

There were some problems with the classification, however. At the time, the atomic weight of beryllium was thought to be 14. If this weight was used, it had to be placed in a group with nitrogen and phosphorus, which had very different properties. Mendeleev boldly concluded that this atomic weight was incorrect. He gave beryllium a weight of 9 instead. This brought it into the magnesium family, where it seemed to belong. Then Mendeleev looked at tellurium, which was supposed to have an atomic weight of 128. That had to be wrong, too. But if it had a weight between 123 and 126, everything would work beautifully; it would then fall into the right group. He made even more dramatic changes in the weights of some elements. For example, at the time uranium was thought to have an atomic weight of 120. This didn't fit into Mendeleev's scheme at all, so he boldly doubled the figure, making it 240 (this is very close to the modern figure, which is 238). He made other changes too. For example, he switched the places of gold and platinum. He had to if his system was to work. He felt fully justified in doing this. It was inconceivable that the appearance of such marvelous order was an illusion.

Mendeleev observed that there were some gaps in his table, empty spaces to which no element was assigned. He concluded that these represented elements that had not yet been discovered. For example, there was a gap under boron, so Mendeleev concluded that it must be an unknown element with properties like boron. He named it eka-boron ("eka" is Sanskrit for the numeral one). Similarly, there were gaps under aluminum and silicon. Mendeleev called these missing elements eka-aluminum and eka-silicon. The positions of the missing elements in his table allowed him to estimate their atomic weights and also to describe their chemical and physical properties accurately.

In 1869 Mendeleev's paper on the periodic law, "On the Relation of the Properties to the Atomic Weights of the Elements," was read before the Russian Chemical Society. Mendeleev was ill at the time and couldn't appear at the Society in person, so a colleague, Nicolai Menshutkin, read the paper. There was little response. If Mendeleev had read it in person, there would have been a question-and-answer period, and he could have cleared up points that puzzled the audience. This might have evoked some interest. However Menshutkin didn't fully understand the paper he had read, and a discussion wasn't possible.

Mendeleev's paper did not evoke much interest either when it was published two months later in the journal of the Chemical Society. However, matters were different when Mendeleev published another paper, "The Natural System of the Elements," two years later. This paper, which contained a revised periodic table (Mendeleev now listed elements in vertical columns rather than horizontal rows), had a much greater response. Russian chemists read it with great interest, as did foreign chemists when the paper was translated into German the same year.

Mendeleev was not alone in making this discovery. In Germany Lothar Meyer discovered the periodic law independently. However, Mendeleev published first and thus had priority. Furthermore, Meyer made no attempt to predict the properties of unknown elements as Mendeleev did. Thus Mendeleev is generally given sole credit for the discovery.

A PREDICTION CONFIRMED

In 1875 the French chemist Emil Lecoq de Boisbaudron discovered Mendeleev's eka-aluminum and named it gallium after the ancient name for France (the right to name an element is the discoverer's even if its existence has already been predicted). When Mendeleev heard of the discovery, he naturally announced that his prediction was confirmed. Lecoq disagreed. Mendeleev had predicted an element with a density of 5.94 (that is, 5.94 times heavier than water) while his element had a density of 4.7. However, when a second determination of the density was made, it was discovered that Lecoq's first result was in error. The actual density of gallium turned out to be 5.91, very close to Mendeleev's prediction.

The discovery of gallium was followed by the discovery of scandium (Mendeleev's eka-boron) in 1879 and of germanium (eka-silicon) in 1886. The new elements had the approximate atomic weights and properties that Mendeleev had predicted. The scientific world was astonished. It is probably safe to say that before Mendeleev's predictions were confirmed, no chemist would have believed that the properties of unknown elements could be predicted with such accuracy.

SEPARATION

At about the same time that Mendeleev's predictions began to be confirmed, his relationship with his wife, Feozva, reached a low point. Because both found the situation unbearable, they decided to separate. It was agreed that Feozva would live on the couple's country estate during the academic year when Mendeleev was in St. Petersburg and that she would live in St. Petersburg when he lived on the estate during the summers. The couple's two children would stay with their mother.

However happy Mendeleev might have been to separate from his wife, he undoubtedly missed his son and daughter. However, it is unlikely that he spent a great deal of time thinking about the matter. He kept as busy as always, and in 1876 the Russian government sent him to study oil-drilling practices in Pennsylvania. The first commercial oil well had been drilled there in 1859, and it was hoped that Mendeleev could use what he learned to make recommendations about the development of Russian oil fields.

While he was in Pennsylvania, Mendeleev visited refineries, interviewed people who worked in the local oil industry, and studied the rock formations in regions where oil was found. Thus in Russia, Mendeleev is remembered not only as a chemist, but also as the father of the Russian oil industry, and justifiably so, because he contributed more than anyone else to the development of Russian oil production. On his return to Russia he invented a new method of refining and tested it under factory conditions. He went to southern Russia to study the oil-bearing land there and wrote a book titled *The Petroleum Industry in Pennsylvania and the Caucasus.*

When Mendeleev traveled around Russia, he didn't do so in the style that most government appointees would have demanded. Instead, he bought third-class tickets so that he could converse with the common people. He learned of their bitter feelings about the repressive Russian government. And these were ordinary people, not political revolutionaries like the exiles he had known in Siberia. From time to time, he denounced the abuses of the Russian bureaucracy. For most people this would have been dangerous, but Mendeleev was Russia's most famous scientist, and there would have been very vociferous protests if he was arrested or taken away by the secret police. Furthermore, he advocated liberal, not revolutionary, ideas. He believed in the possibility of reform and never advocated over-throwing the monarchy. So the tsarist government employed the strategy of sending him away on some government mission when-ever his complaints caused too much embarrassment.

Mendeleev did not share the attitudes toward women that were common in his day. He didn't believe that women were fully the intellectual equals of men, but he thought that men and women should be treated equally in the workplace and that women should have the same educational opportunities as men. Consequently he admitted women to his lectures at the university, something that was almost unheard of in his day.

His liberalism made him some enemies but because of his stature they were unable to do him much harm. In 1880 the Imperial Academy of Sciences refused to elect Mendeleev to membership, electing instead Friedrich Beilstein, a German professor at the Impe-rial Technological Institute. But the University of Moscow soon made him an honorary member, possibly in response to the Imperial Academy's action. Mendeleev seems not to have been bothered by the snub. In any case, he soon had honors enough, receiving the Davy medal from England's Royal Society and the Faraday medal from the English Chemical Society. When Mendeleev came to England to receive these honors, the English called him "Faust"; he was the magician who had predicted the properties of elements that no one had ever seen. He received numerous other honors also, including awards and honorary degrees from the German and American chemical societies and from the universities of Princeton, Cambridge, Oxford, and Göttingen.

MENDELEEV THE BIGAMIST

In 1887 Mendeleev's sister Ekaterina came to St. Petersburg to keep house for him. She brought her children with her and also a 19-year-old girl, Anna Ivanova Popov, who studied at the St. Petersburg Academy of Art with Ekaterina's daughter. At first Mendeleev saw little of either Ekaterina or Anna. The part of the apartment that contained Mendeleev's bedroom and study had a private entrance. Normally he didn't see any of the other occupants of his apartment except when Ekaterina brought him his meals. However he soon began to catch glimpses of an attractive young woman. Before long he found himself becoming infatuated with her. Naturally this presented difficulties. After all, he was still married.

Mendeleev married Anna in 1882 after divorcing his wife. Russian law prohibited remarriage for seven years after a divorce. But Mendeleev paid an Orthodox priest, who was later defrocked, to give him a dispensation. When a member of the tsar's court remarked on Mendeleev's bigamy, the tsar is supposed to have replied, "I admit Mendeleev has two wives, but I have only one Mendeleev." Mendeleev's conduct was subsequently ignored.

The marriage was happy, and being married to Anna changed the character of Mendeleev's life. He had previously spent most of his time alone. Now, twice a week, he and Anna held informal parties. The guests were artists, musicians, and scientists. Mendeleev became interested in art, and he even wrote an article about a painting for a newspaper. In 1894 the Academy of Art elected him to membership. Ironically, he had gained admission to the Academy of Art but not to the Academy of Sciences.

PROTEST

When Alexander II was assassinated in 1881, Alexander III, who succeeded him, continued his father's harsh policies. In some respects the son's government became more oppressive than his father's. This was certainly true in the case of education. Count I. D. Delyanov, who was appointed minister of education, lowered the quota for Jewish students, so that many of them could not obtain any education above the elementary level. Opposed to education for women, he closed the Women's Medical College, an institution founded by Borodin. Wishing to deny higher education to the "children of coachmen, footmen, laundresses and small shopkeepers," Delyanov eliminated many of the government scholarships like the one that had allowed Mendeleev to obtain an education, and he raised tuition fees. In defiance of the government, Mendeleev continued to admit women to his lectures. Again he was left alone. Thinking of himself as "an evolutionist of a peaceful type," he did not sympathize with the revolutionary ideas of some of his students. Nevertheless, he used his influence to help them when they got into trouble.

One day in March 1890 Mendeleev encountered a large student demonstration as he left the university. The students were agitating for the reversal of Delyanov's edicts. Later that day, one of his chemistry students came to his apartment and asked him to come to a student meeting that was to be held at a later date. Mendeleev went. The main order of business was the reading of a petition that a student committee had prepared. It was immediately accepted. But this raised the question of who should take the petition to Delyanov. If a student did this, Delyanov was likely to throw it into a wastebasket without reading it so Mendeleev was asked to present the petition. He agreed.

Mendeleev took the petition to Delyanov's office two days later. Some time after that a messenger brought the unopened envelope back. The unread petition was accompanied by a message that read: "On the instruction of the Minister of Education, the enclosed document is returned to Councilor of State Professor Mendeleev, since neither the Minister nor anyone else in the service of His Imperial Majesty has the right to accept a document of this nature." Mendeleev was furious. The next day he submitted his resignation to the university. Neither his colleagues nor the rector were able to persuade him to reconsider nor was he swayed by his students' pleas. Mendeleev gave his last lecture on March 22.

Mendeleev and Anna had to find a new apartment, because the university provided the one they were living in. When they had moved in, Mendeleev set up a laboratory in one of the rooms. This allowed him to continue to perform experiments and to submit papers to scientific journals.

Although Mendeleev's resignation was an act of political protest, the Russian government continued to consult him on various matters. He was far too valuable a resource to be ignored. For example, like Lavoisier, he was put to work improving gunpowder. Receiving a commission from the Russian admiralty to improve the smokeless powder that was then in use, Mendeleev set to work at once and within a year he produced a product that was superior to most foreign powders.

BUREAU OF WEIGHTS AND MEASURES

In 1893 the Russian minister of finance, who was familiar with Mendeleev's contributions to Russian industry, offered him the post of director of the Russian Bureau of Weights and Measures. Mendeleev accepted, and he and his family (he and Anna now had several children) moved once again, to an apartment in one of the bureau's buildings. Mendeleev launched into the job with enthusiasm.

Determined to bring order to the chaotic systems of measurement then used in Russia, he used techniques developed abroad and invented some measuring devices himself. He established the metric system in Russia and insisted on a greater precision in measuring equipment.

When Mendeleev discovered that his employees had been given substandard housing, he badgered government officials until new apartments were built for them. He employed women as well as men in the Bureau and found that some of the women did work that was superior to that of their male counterparts.

In 1899 the Russian government asked Mendeleev to go to the Urals to study the iron industry. He remained there until he had enough material for an 866-page book, *The Iron Industry of the Urals in the Year 1899*. In 1902 he revised his periodic table, adding a new column for the newly discovered inert gasses (helium, argon, krypton, and xenon). He also became a map maker, producing a large map of Russia that was more accurate than any that preceded it. Though he was now nearly 70, it seemed that Mendeleev had more energy than ever.

In 1902 Mendeleev was forced to take a short rest from his labors. Like his father, he developed cataracts and had to have them operated upon. His eyes were bandaged for two weeks. When the bandages came off and he found that his vision had improved, he plunged back into his work.

REVOLUTION

In January 1905 St. Petersburg experienced a series of strikes, and the organization that had fomented some of them, the Assembly of Russian Workingmen, decided to present a request for reforms to Tsar Nicholas II. The Assembly's leader, a monk named Georgy Gapon, arranged a demonstration before the tsar's winter palace. But the tsar was away at the time, and matters fell into the hands of his uncle,

Grand Duke Vladimir. Though it was a peaceful demonstration, the duke ordered the police to open fire on the demonstrators. More than a hundred were killed and hundreds of others were wounded.

The massacre was followed by general strikes in St. Petersburg and other cities and by peasant uprisings. Military units, including army units stationed along the Trans-Siberian Railroad, also joined the revolt. Nicholas attempted to quell the revolt by announcing the formation of an elected assembly to serve in an advisory capacity to the government. But when election procedures were announced in August, there was even more protest, and a railroad strike that began in October incited general strikes in most of the large cities. But in the end, the revolution failed. The government arrested most of the revolutionary leaders, and the military suppressed revolts in Georgia, in the provinces bordering the Baltic Sea, and in Poland. The government regained control of the Trans-Siberian Railroad and quelled revolt in rebellious army units.

Although the revolution failed to overthrow the tsarist government and set up a democratic state in its place, some of its goals were achieved. The government felt compelled to institute numerous reforms, and it created a legislative body, the Duma. The hope cherished by Mendeleev and other liberals that change could be brought about by peaceful evolution had been dashed. However it did appear that Russia was beginning to evolve into a democracy.

Mendeleev died too soon to witness the far more violent revolution in 1917. In the fall of 1906 he fell ill. He was diagnosed with influenza, and he went to Cannes in southern France to recuperate. At first he seemed to have recovered, but symptoms of his illness began to reappear after he returned to Russia. Matters became worse in January 1907 when he contracted pneumonia, and during the early hours of January 20 he died.

Mendeleev was buried in the Volkovo Cemetery on January 25. His funeral was attended by thousands, including some students carrying a large tablet on which the periodic table was inscribed. That night the streetlights of St. Petersburg were draped with black crepe.

In 1955 Albert Ghiorso and his colleagues at the University of California at Berkeley discovered the artificial element mendelevium. The scientists produced mendelevium one atom at a time, getting 17 atoms in all. Mendelevium was added to the periodic table as element number 101.

CHAPTER 10

DECIPHERING THE ATOM

A s the nineteenth century entered its final decade, chemists still had no hint as to why there should be so many different chemical elements. However, few of them pondered the mystery. The basic principles of chemistry had been established, and Mendeleev had shown that there was order in the table of the elements. There were still unanswered questions, as there always are in any scientific field, but chemistry no longer seemed to be beset by contradictions or by fundamental unsolved problems.

Thus it was the physicists who took the next steps toward understanding the nature of matter. In 1896 the French physicist Henri Becquerel discovered radioactivity, and in 1897 the English physicist J. J. Thomson* discovered the first subatomic particle, the electron. Subsequently, studies of the radiation emitted by radioactive atoms showed that these atoms emitted "radiation" of three different kinds, which were called alpha, beta, and gamma after the first three letters

---

*His full name was Joseph John Thomson, but everyone called him "J.J."

of the Greek alphabet. It was soon found that only the third was really radiation. Alpha "rays" turned out to be made up of particles identical to the nucleus of the helium atom, while beta radiation was made up of electrons.

Clearly, atoms were not the indivisible things they were always thought to be. Chemistry had made great steps toward deciphering the key to the universe, but there was much more to be learned. If physicists were to learn what the universe was made of, it would be necessary to understand atomic structure.

NIELS BOHR

Bohr's father, the physiologist Christian Bohr, was the first in the family to get a Ph.D. and to pursue a university teaching career. He was a distinguished scientist who was recommended for the Nobel Prize in 1907 and 1908 for his work on the release of oxygen by hemoglobin. In 1875 Danish women were permitted to pursue university studies for the first time, and Christian, who was then a medical student, took the job of coaching some to prepare them for admission. One of his students was a young woman named Ellen Adler, whom he married in 1881. On her 25th birthday, October 7, 1885, Ellen gave birth to a son, whom the couple named Niels Henrik David Bohr.

By the time he entered the University of Copenhagen in 1903, Niels Bohr had become very interested in physics and mathematics so he chose physics as his major subject. But his interests were not limited to scientific subjects. Both Niels and his younger brother, Harald, had been playing soccer since high school and they continued to do so when they were university students. Bohr's biographers sometimes say that both Bohr brothers became famous in Denmark as soccer players. However, Harald was by far the better known, playing in the Olympic games in 1908, when the Danish team won the silver medal. In 1955 Niels said of his brother in a taped conversation, "He really was a famous man—all that nonsense that I was a great soccer player is very dubious."

Possibly Niels didn't play soccer with the concentration that Harald did. Once, when his team was playing a German soccer club, most of the activity was on the German side of the field. At one point, the ball came rolling toward the Danish goal. Niels was the goalie, but instead of running to grab the ball he remained standing where he was, contemplating the goalpost. It was only when a spectator shouted at him that he sprang into action. After the game, the embarrassed Bohr admitted that he had become absorbed in a mathematical problem and had been looking at some calculations he had performed on the goal post.

In 1910 Niels began work on his Ph.D. degree. Harald, the younger brother, received his doctoral degree the same year. In those days, relatively few Ph.D.s were given, and a candidate's defense of his dissertation was a public event that was announced in the newspapers. The defenses weren't the relatively informal affairs that they are now; the candidate was required to appear in white tie and tails. When Harald—who went on to become a noted mathematician—defended his dissertation, most of those attending were soccer players, and a Danish newspaper reported that the soccer player Harald Bohr had become "a rising comet in the heavens of mathematics."

Niels defended his dissertation the following year. According to the newspaper reports the next day, the auditorium in which the defense took place was filled, and some of the spectators had to stand in the corridor (which suggests that perhaps Bohr was better known than he claimed). The dissertation was printed as a book shortly afterwards, and Bohr sent copies to a number of notable physicists but none of them responded. The most likely cause was that the book was published in Danish. If it were in German or French, they could have read it, or read it more easily.

ENGLAND

In 1911 Bohr went to Cambridge, England, to do postdoctoral research at the Cavendish laboratory under J. J. Thomson. At the time, the Cavendish was one of the two leading centers for research in physics (the other was the Physico-Technical Institute in Berlin). Both Thomson and Bohr were deeply interested in electron theory. Thomson had discovered the particle, and Bohr's dissertation dealt with the behavior of electrons in metals. While doing his dissertation research he made a discovery that cast doubt on one of Thomson's ideas. He hoped that he would be able to engage in fruitful discussions about the matter with Thomson.

Unfortunately, Bohr's first meeting with the English physicist did not go well. He walked into the meeting carrying one of Thomson's books, opened it to a certain page, and said, "This is wrong." Most likely he would not have been so blunt if his English was better. However, at this time Bohr's command of English was relatively poor. Later in life he wrote, "It was a disappointment that Thomson was not interested to learn that his calculations were not correct. That was also my fault. I had no great knowledge of English and therefore I did not know how to express myself. And I could only say that this is incorrect." Bohr's relationship with Thomson quickly became strained, and Thomson began to avoid Bohr when he saw him in the laboratory. Bohr soon began to feel that the time he was spending at the laboratory was fruitless. Apparently, he had gained little by coming to England. But then, in the fall of 1911, he met Ernest Rutherford.

Bohr went to Manchester to visit a friend of his late father and arranged to meet Rutherford, who maintained a laboratory there. Bohr told Rutherford that he would like to come to Manchester to work with him, and his transfer to Rutherford's laboratory was soon arranged. At first Rutherford put Bohr to work doing experimental work, but when he expressed an interest in doing theory, he was relieved of the obligation to perform experiments.

A RABBIT FROM THE ANTIPODES

Ernest Rutherford was born in 1871, the fourth of 12 children, on a farm on the South Island of New Zealand, a predominantly rural environment. At the time, the South Island had a population of about 250,000, and the British had settled it only 50 years earlier. Rutherford's parents, a Scottish wheelwright named James Rutherford and his wife Martha, who had been an English schoolteacher before the two emigrated to New Zealand, believed in the value of education and made sacrifices so that their children could be educated.

In 1887 Rutherford won a scholarship to nearby Nelson College, a secondary school. Another scholarship brought him to Canterbury College in Christchurch, where he got a B.A. in 1892 and an M.A. in 1893. He stayed at the college an extra year to do research in physics, supporting himself by teaching part time. The papers he published on his research won him yet another scholarship, this time to Cambridge University in England. At Cambridge, Rutherford worked under J. J. Thomson in the Cavendish laboratory, continuing the research that he had done in New Zealand. His zeal seems to have impressed his colleagues, especially the younger ones, one of whom remarked, "We've got a rabbit from the Antipodes and he's burrowing mighty deep."

When the French physicist Henri Becquerel discovered radioactivity in 1896, Rutherford immediately became interested in the phenomenon, and he published a paper on "uranium rays" in September 1898. By the time the paper appeared, however, Rutherford was on his way to Canada, where he had an appointment as professor of physics at McGill University in Montreal. He might have stayed on at Cambridge except that he had planned to marry as soon as possible ever since he left New Zealand. Before leaving, he had become engaged to Mary Newton, the daughter of his landlady. After arriving in Canada, he wrote to his fiancée, "The salary is only 500 pounds but enough for you and me to start on." In 1900 Rutherford went to New Zealand to visit his parents and to get married. When their only child, a daughter, was born in 1901, Rutherford wrote to his mother, "It is suggested that I call her 'Ione' after my respect for ions in gases." But in the end he and Mary settled on the name Eileen.

At McGill, Rutherford continued to study the radiation emitted by uranium, and he soon established that there were two distinct types, which he called alpha and beta after the first two letters of the Greek alphabet. In 1902 Rutherford and the English chemist Frederick Soddy, who was also at McGill, established that radioactivity was a process by which atoms of one element disintegrated into atoms of another. At first, physicists thought that the alpha particles were electrically neutral. But Rutherford showed in 1903 that they were electrically charged. In 1908, after a decade of experiments on radioactivity, Rutherford finally established that alpha particles were positively charged helium atoms.

In the fall of 1907 Rutherford went back to England, this time to the University of Manchester, where he had been appointed professor of physics and director of the physics laboratory. By this time he was probably the world's leading expert on radioactivity. He had published some 80 scientific papers while at McGill. Back in 1902 he had written to his mother, "I have to keep going, as there are always people on my track. I have to publish my present work as rapidly as possible in order to keep in the race. The best sprinters in this road of investigation are Becquerel and the Curies in Paris, who have done a great deal of very important work on the subject of radioactive bodies during the last few years." Perhaps it seems a little arrogant of Rutherford to compare himself to Becquerel and the Curies, who shared the Nobel Prize in physics the following year. However, he understood the significance of the research he was doing.

In 1908 Rutherford received the Nobel Prize himself, but not for physics. He was given the prize for chemistry "for his investigations into the disintegration of the elements, and the chemistry of radioactive substances." In his speech at the Nobel Prize awards banquet, Rutherford remarked that he "had dealt with many different transformations with different time periods, but the quickest he had met was his own transformation from a physicist into a chemist."

DISCOVERY OF THE ATOMIC NUCLEUS

Rutherford was the first physicist to make his greatest discovery *after* receiving the Nobel Prize. In 1911 he discovered that atoms were not what physicists thought them to be. In doing so, he took the first step toward unraveling the secrets of atomic structure. Before 1911, physicists generally adhered to the "plum pudding" model of the atom, which had been devised by Thomson. According to Thomson's theory, the raisins—the negatively charged electrons—(a plum pudding actually contains raisins and currants, not plums) vibrated back and forth through a sphere of positive electricity (the pudding). The theory seemed plausible, because vibrating electrons could emit light and other forms of radiation. However, it had never been tested experimentally.

In 1910 Rutherford wrote to a friend, "I think I can devise an atom much superior to J.J.'s, for the explanation of and stoppage of alpha and beta particles, and at the same time I think it will fit in extraordinary well with the experimental numbers." Rather than devise a model of the atom based on theoretical ideas as Thomson had done, Rutherford intended to probe atomic structure by bombarding atoms with particles ejected from radioactive atoms. Rutherford felt that experimental physics was the only *real* physics and that by performing experiments he could gain greater insight into atomic structure than Thomson had been able to get using only theory.

Rutherford decided that the best way to proceed was to bombard the atoms with rapidly moving alpha particles. If Thomson's theory was correct, the paths of the particles would be deviated only slightly. Electrons were thousands of times lighter than alpha particles; they couldn't possibly deflect the paths of the alpha particles significantly. Neither could the hypothetical sphere of positive electricity that Thomson had postulated; it was too diffuse. So Rutherford and his assistants directed the particles at sheets of thin gold foil, approximately 0.00004 centimeters thick. After the particles passed through the foil, they struck a fluorescent screen. The impacts produced tiny flashes of light that could be observed through a microscope.

Rutherford didn't actually make any observations himself. His assistants Ernest Marsden and Hans Geiger (who later invented the Geiger counter), who were Cambridge students at the time, made them. One day Rutherford came into the room where they were counting flashes and said, as Marsden later recalled, "See if you can get some effect of alpha particles directly reflected from a metal surface." Marsden and Geiger had been counting particles that were deviated by small angles. Rutherford was suggesting that they should try to see if any of the particles were reflected at large ones. The assistants found that although the deflection angle of most of the alpha particles was slight, a small number were deflected at large angles, and about 1 in 8,000 was deflected at an angle of 90 degrees or more. "It was almost as incredible as if you fired a 15-inch shell at a piece of tissue paper, and it came back to hit you," Rutherford said later. At the time, England and Germany were engaged in a naval arms race and this could very well have influenced his choice of metaphor. However, the result was very surprising. The alpha particles traveled at velocities of thousands of miles per second toward supposedly insubstantial atoms, yet some of them bounced right back.

Rutherford was often disdainful about theoretical physics, sometimes excessively so. For example, in 1910 the German physicist Wilhelm Wien, who was visiting England, expounded Einstein's relatively new theory of relativity to Rutherford. After going into certain points relating to the theory, Wien remarked that no Anglo-Saxon could understand that. Rutherford laughed and replied that, no, they had too much sense. But he could perform theoretical calculations when he had to, and he quickly set to work trying to gain an understanding of the phenomenon that Marsden and Geiger had observed. When he worked out some mathematical equations describing the paths of the alpha particles, he found that the result could be explained in only one way. The positive charge of an atom was not distributed over a sphere that was about the size of the atom itself. It was confined to a tiny nucleus that was much smaller than the atom. If an alpha particle passed very close to the nucleus, then electrical repulsion caused it to be deflected through a large angle.

Rutherford's discovery of the atomic nucleus was his greatest contribution to physics and it established him as the leading experimental physicist of his day. However, it was only a beginning, and many questions about the atom remained unanswered. As yet nothing was known about electron orbits or about the relationship between the structure of the atom and the periodic table. Before Rutherford performed his experiments, it was thought that the atom was understood. Now it was apparent that much remained to be learned. But then great discoveries in physics seem always to suggest new questions and open up new lines of research. The more that is known, the better the picture scientists have of what remains unknown.

Rutherford was knighted in 1914, and during World War I he made the Manchester laboratory into a center for research on defense against submarine attacks. His former assistants, Marsden and Geiger, didn't participate in the research, however. They were both on the Western front, on opposing sides.

In 1919 Rutherford succeeded Thomson as director of Cambridge's Cavendish laboratory, and in that position did less research. However, the experimental work that was done by others under his direction was significant (some of it will be discussed in the next chapter). Rutherford had a positive influence on numerous young scientists, and the Cavendish laboratory remained as important a center for physics research as it had been under Thomson.

In 1931 Rutherford was elevated to the peerage, becoming Ernest, Lord Rutherford of Nelson. He put a kiwi and a Maori warrior holding a club on his coat of arms and also a figure representing Hermes Trismegistus, the mythical founder of alchemy. However, this was not a happy time for Rutherford. His only daughter had died at the age of 29 in December of the previous year. As he grew older, Rutherford's life fell into a pattern that is common among well-known older scientists: serving on government commissions. When Hitler rose to power in Germany, Rutherford helped found the Academic Assistance Council, which aided displaced academics, and later accepted the post of president of that organization.

Rutherford died of a strangulated hernia on October 19, 1937, at the age of 66. On October 21 his remains were cremated. His ashes were interred in Westminster Abbey among the remains of England's other great scientists, including Newton. When J. J. Thomson died in 1940, his remains were placed next to those of Rutherford, the man who had disproved his theory of the atom.

ATOMIC STRUCTURE

Shortly after coming to Rutherford's laboratory, Bohr set to work on the problem of understanding the structure of atoms. Rutherford's discovery of the atomic nucleus had introduced formidable problems. It seemed necessary to assume that the electrons in an atom orbited the nucleus. Otherwise, the electrical attraction between the electrons and the nucleus would cause the electrons and the nucleus to collide with one another. But, as we have seen, the assumption that the electrons orbited the nucleus didn't seem to work either. Orbiting electrons should lose energy and fall into the nucleus anyway.

When physicists are confronted with new problems, they almost always consider the simplest cases first. The more complicated ones can always be tackled later. So Bohr began by considering the hydrogen atom, which contained just one negatively charged electron and a nucleus with a positive charge. Bohr realized that there was an important clue concerning the behavior of the electron. The light emitted by bodies that were heated had to come from their constituent atoms; there was nothing in them but atoms, after all. Furthermore, the emission of light had to have something to do with the motion of the electrons. Electrons had differing amounts of energy depending on their distance from the nucleus. Presumably one that suddenly moved closer to the nucleus could give up some of its energy by emitting light.

There was another clue, too. In 1901 the German physicist Max Planck had propounded his quantum theory and shown that atoms emit light in certain discrete quantities, or quanta. An atom could emit one quantum of light, or two, or six, or any other whole number. But it couldn't emit one and a half quanta, or three and a third, or any fractional number. Bohr realized that Planck's result could be explained if the electron in a hydrogen atom could revolve around the nucleus only in certain prescribed orbits with definite energies. It could follow these orbits, but not any in between. An atom emitted light when an electron suddenly made a jump from one orbit to another.

One problem remained, that of explaining why the orbiting electrons didn't lose energy and fall into the nucleus. Bohr solved the problem by making the assumption that the laws of electricity and magnetism that said this should happen simply didn't apply to events on the subatomic level. This was the most audacious part of his theory. Never before had any physicist been willing to assume that the known laws of physics were inoperable under certain circumstances.

THE HYDROGEN ATOM

Although Bohr had begun work on his quantum theory of the hydrogen atom while he was still in Manchester, he didn't publish his paper on the subject until 1913, when he was back in Copenhagen. He left England in July 1912 and soon got a position teaching physics to medical students at the University of Copenhagen. After getting married on August 1, 1912, Bohr and his wife, Margrethe, went to England for a honeymoon. On their return in September, Bohr returned to the problem of deciphering the hydrogen atom. In March 1913 he sent his first paper on the subject to Rutherford, asking him to send it to *Philosophical Magazine* for publication.* Rutherford had

---

*In those days scientific papers could not be published unless recommended by a well-known scientist. Thus Bohr could not submit the paper directly.

some reservations about the paper. For example, he asked Bohr how the electron knew what orbit it should jump to. However, he sent the paper on and it was duly published.

The initial reaction to Bohr's theory was mixed. Just as his paper appeared, some physicists were meeting at a conference in Göttingen in Germany. One of those attending was Bohr's brother Harald, by now a well-known mathematician. Harald reported back that the German physicists were in agreement that "the whole thing was some awful nonsense, bordering on fraud." Einstein didn't like the theory very much either when he first heard about it. In a letter to another physicist, he expressed the opinion that if such a crazy theory proved to be correct, then physics would be at an end; it would no longer be possible to do physics. The German physicists Max von Laue and Otto Stern had a similar reaction. They vowed to one another that if this crazy theory turned out to be right, then they would leave physics (but when it did turn out to be correct, they didn't).

But there were others who viewed Bohr's work more favorably. For example, the German physicist Arnold Sommerfeld, who was later to elaborate on Bohr's theory, described Bohr's contribution as "an extremely important paper ... which will be a milestone in theoretical physics." But in the end, Bohr's theory triumphed because it worked. It made predictions that could be tested by experiment. For example, it accurately predicted what wavelengths of light a hydrogen atom should emit. When an electrical discharge is passed through a gas, light is emitted. But the entire spectrum is not seen. Instead, the hydrogen atoms emit light of certain specific wavelengths that correspond to the energy given off in specific quantum jumps.

Bohr next applied his theory to helium ions—helium atoms in which one of the two electrons is removed—and again the predictions of the theory exactly matched results obtained in experiments. The scientific world was convinced. For example, when Einstein heard of the results, he reversed himself and said, "This is a tremendous achievement—Bohr's theory must be right."

In 1914 Rutherford offered Bohr a position in Manchester, which he accepted. By now Bohr's fame was rapidly growing, and officials at the University of Copenhagen began to fear that he might not return to Denmark. Thus they sought to obtain a full professorship for him. Initially the Danish government denied their request and only in 1916 was the appointment approved. Conditions were attached, however. Bohr was to continue to teach medical students in addition to serving as a professor.

Performing two jobs would hardly have allowed Bohr much time for research. But arrangements were made according to which H. M. Hansen, the lecturer who had taught the medical students in Bohr's absence, would continue to do so without being paid by the university. Instead, he received a salary from money provided by outside sources. It was only in 1918 that the university made Hansen a salaried employee. Hansen eventually became rector of the university, serving in that position for many years.

In 1916 it was the custom that a new professor, attired in morning coat and white gloves, present himself before the Danish king and queen. During the audience, Christian X remarked to Bohr that he was delighted to meet the famous soccer player. Bohr replied that it was his brother who was the famous one. This was a breach of etiquette; one wasn't supposed to contradict the king during an audience. So Christian repeated the remark, expecting Bohr to answer in a more suitable manner. Bohr said that, yes, he was a soccer player, but it was his brother who was the famous one. At this point Christian terminated the audience, and Bohr left, walking backwards as was customary. One didn't turn one's back on Danish royalty.

THE INSTITUTE FOR THEORETICAL PHYSICS

In 1917 Bohr applied to the faculty of natural science for funding to establish an institute for theoretical physics. What he had in mind was a laboratory in which he and his students could perform experiments. This involved no contradiction. In 1917 the distinction between theoretical and experimental physics was just beginning to be made, and the term "theoretical physics" was often used to mean fundamental physics.

Meanwhile an old friend of Bohr's, Aage Berlème, began a campaign to raise money from private sources to buy a site for the institute. The campaign reached its goal in 1917, and construction began in 1918. Private contributions continued, allowing Bohr to buy equipment for the institute. The largest donation came from the Carlsberg Foundation, established by the makers of Carlsberg beer. The foundation continued to make grants to the institute, and by the time Bohr died in 1962, it had made more than a hundred.

By the end of the 1920s, Bohr's institute (the official name was the University of Copenhagen Institute of Theoretical Physics, but everyone called it "Bohr's institute" instead) had become the world's most famous center for research in physics. It was visited at one time or another by many of the most notable physicists of the first half of the twentieth century, and many noted physicists of the second half of the century did postdoctoral work there. Physicists often visited the institute to have discussions with Bohr, who was recognized as a leading physicist within a few years of the publication of his quantum theory of the hydrogen atom. And in 1922 he was awarded the Nobel Prize. Bohr and Einstein received the prize the same year. But it wasn't a shared prize; Einstein's was for 1921.

The great respect that his colleagues had for Bohr is shown by the following anecdote: In 1925 two young unknown Dutch physicists, Samuel Goudsmit and George Uhlenbeck, advanced the hypothesis that electrons had spin. When their paper was published, physicists weren't quite sure what to make of it. This wasn't surprising because Goudsmit and Uhlenbeck had some doubts themselves. After giving their paper to physicist Paul Ehrenfest and asking him to send it to a journal, they tried to withdraw it. But Ehrenfest replied that he had already sent the paper off for publication, adding that the two young physicists were young enough to make fools of themselves. What Ehrenfest meant was that if the idea turned out to be wrong, it wouldn't harm their reputations because they didn't yet have reputations.

It so happened that in December 1925 Bohr took a train to Lieden to attend ceremonies honoring the Dutch physicist Hendrik Lorentz. When the train stopped in Hamburg, it was met by Otto Stern and the Austrian physicist Wolfgang Pauli, who had come to ask Bohr what he thought about spin. When Bohr arrived in Leiden, he was met by Ehrenfest and Einstein, who asked him what he thought about spin. After leaving Leiden, Bohr went to Göttingen. He was met there by the German physicists Werner Heisenberg and Pascual Jordan, who asked him what he thought about spin. When he was on his way home, his train stopped in Berlin, where it was met again by Pauli, who asked Bohr what he thought about spin *now*. By this time Bohr had decided that he favored the idea, while Pauli remained skeptical, calling the idea "another Copenhagen heresy." But in the end Bohr turned out to be right. Electron spin was indeed something real.

Discussing physics with Bohr was sometimes quite exhausting. In 1925, while working at Bohr's institute, Heisenberg discovered quantum mechanics, the theory that superseded the "old quantum theory" that had been developed by Bohr and his colleagues. Then, in 1926, a theory that looked much different from Heisenberg's, but which turned out to be mathematically equivalent, was propounded by the Austrian physicist Erwin Schrödinger.

Bohr promptly invited Schrödinger to Copenhagen to discuss quantum mechanics with him and Heisenberg. The discussions began when Schrödinger arrived at the Copenhagen train station and went on for several days, usually ending late at night. They were so intense that Schrödinger finally became ill, probably from exhaustion, and took to his bed. However, Bohr felt that there were still some points to be settled, and he was seen at Schrödinger's bedside, continuing the discussion of quantum mechanics.

THE PERIODIC TABLE

In 1920 Bohr turned his attention to the problem of atomic structure. Matters had become somewhat more complicated than they were in Mendeleev's day. By 1920, 14 elements had been discovered that did not seem to follow Mendeleev's periodic law. Called the rare earths, they had similar properties and followed one another in the table of elements; they were elements 58 through 71. When Mendeleev formulated his law only two had been discovered, so they didn't seem to present any great problem. But now they presented an anomaly that no one had been able to clear up. A workable theory of atomic structure would have to explain not only why periodicities were seen in the larger part of the table of the elements but also why they disappeared when one came to the rare earths.

By then X-ray experiments allowed scientists to determine the charges of atomic nuclei and, because atoms were electrically neutral, find out the number of electrons in an atom of any element. For example, if the charge on a nucleus was +27, there had to be 27 electrons in the atom to balance that out. In 1916 the German physicist Walther Kossel had speculated that electrons in atoms arranged themselves into concentric shells. For example, argon, which had 18 electrons, had 2 in the innermost shell, 8 in a second shall that surrounded it, and 8 more in a third. But Kossel could not explain why this should be, and he considered no atoms with more than 27 electrons.

By the time Bohr turned his attention to the problem, significant advances had been made. Physicists working with the old quantum theory had developed a number of rules about the manner in which electrons interacted with one another. Bohr realized that these rules could be used to confirm Kossel's hypothesis and to make informed guesses about the atomic structure of the elements. For example, hydrogen has one electron, placed in the innermost shell. Helium, having two electrons, has this shell filled up. Thus lithium, the third element, has to have two electrons in an inner shell and one with an orbit outside the shell. With each successive atom, one electron is added, until you get to neon, with two electrons in the first shell and eight in the second. Because the second shell is now full, the next element, sodium, must have one electron in a third shell.

But matters do not always proceed in so orderly a manner. Some elements have an electron in an outer shell when an inner shell is not completely full. This is what happened in the case of the rare earths. They all had the same number of electrons in the outermost shell, giving them similar physical and chemical properties. But except for the last rare earth there were gaps in the inner shells.

Bohr's theory not only solved the puzzle of the rare earths, it also explained why Mendeleev's periodic law works so well in most cases. Elements with the same number of electrons in the outermost shell have similar properties. For example, both sodium and potassium have a single electron in the outer shell. Magnesium and calcium, which are also chemically and physically similar, each have two. Carbon and silicon have four. And so on.

When Bohr published his first paper on the topic in 1921, the physicists who read it were convinced that his results were based on undisclosed calculations. They didn't see how so complex a theory could be worked out without making use of some mathematical foundation. But they were wrong. Bohr often proceeded intuitively, using whatever principle seemed most appropriate, as he considered one or another of the elements. Given his methods, it isn't surprising that Bohr made some faulty assignments. Nevertheless, his picture of atomic structure is basically the same as the one used by chemists and physicists today.

Bohr's theory received a striking confirmation when the element hafnium was discovered at his institute in 1923. During the early 1920s most chemists believed that element 72 would turn out to be a rare earth. But Bohr's theory implied that this element should have four electrons in its outermost shell, not three as the rare earths did. It should therefore have properties similar to those of the element zirconium.

In 1923 the French chemist Georges Urbain claimed to have discovered element 72 and to have established that it was indeed a rare earth, which he named celtium. But Bohr was skeptical. When he discussed the matter with the Dutch experimental physicist Dirk Coster, he found that Coster didn't believe Urbain's claims either. So Coster and the German physicist Georg von Hevesy, who was also working at the institute, borrowed some mineral samples containing zirconium from the Mineralogy Museum in Copenhagen. They reasoned that if element 72 did indeed have properties like those of zirconium, traces of it were likely to be found in the samples. Coster and von Hevesy found the sought-for element in all of the samples. Not only wasn't it a rare earth, it wasn't especially rare. It later proved to be as common as tin. They named the new element hafnium.

QUANTUM MECHANICS

Volumes have been written about Bohr's role in the development of an interpretation of quantum mechanics and about his many-years-long series of arguments with Einstein on the subject. Because a reasonably complete discussion of the topic would easily double the length of this chapter, I will pass over it briefly, and because a brief discussion generally leaves a lot of questions unanswered, I urge any reader who is especially interested in the subject to consult one of the numerous books that discuss it at greater length.*

Quantum mechanics, which replaced the old quantum theory, was not easy to interpret. It conceived of both light and particles as having a dual nature. They were sometimes observed to be waves and sometimes particles, depending on the type of experiment that one performed. For example, the electron seemed sometimes to be a particle and sometimes a packet of waves. Furthermore, quantum mechanics described the subatomic world in terms of probabilities.

---

*For example, my book, *The Big Questions* (Henry Holt, 2002).

Such phenomena as the emission of a quantum of light from an atom or the emission of an alpha particle from a nucleus were random events. One could speak only of the probability that they would occur within some given period. It was not possible to predict when such an event would take place.

Einstein refused to accept the idea of such randomness in nature. "*Gott würfelt nicht*" ("God doesn't play dice"), he insisted more than once. So he invented a series of "thought experiments" in an attempt to show that quantum mechanics was an inconsistent theory. (A thought experiment is an experiment that cannot actually be performed, but which nevertheless can be used to elucidate some fundamental principle.)

Einstein's arguments with Bohr began in 1927 and continued until 1935. Again and again, Einstein argued against Bohr's interpretation of quantum mechanics, and each time Bohr refuted him. The argument was never really settled during the lifetimes of the two men, however. It was only after both men were dead that new theoretical and experimental work showed that Einstein had been wrong and Bohr right.

NUCLEAR FISSION

During the 1930s Bohr became interested in nuclear physics, and he did some groundbreaking theoretical work in that area. Thus he was very intrigued when he learned that the German physicists Otto Hahn and Lise Meitner had discovered nuclear fission in 1938.* Their paper on this discovery was published in January 1939. By June of that year Bohr and the young American physicist John Archibald Wheeler had completed a paper on the subject. Making use of some of Bohr's previous theoretical work on nuclear physics, they explained how fission took place and showed that uranium 235 was a highly fission-

---

*Many people feel that Meitner should have shared Hahn's Nobel Prize, which he was awarded for this discovery.

able material. Bohr immediately realized that it was theoretically possible to use uranium 235 to make a bomb if enough of it could be separated from the more common uranium 238.

One day as the Italian physicist Enrico Fermi and George Uhlenbeck (who had come to the United States on a visit) were looking out a window overlooking Manhattan, Fermi remarked, "You realize, George, that one small fission bomb could destroy most of what we see outside?" Fermi was soon to be doing some of the preliminary experimental work that preceded the American atomic bomb project. It was Fermi who produced the first controlled nuclear chain reaction.

It was largely because of Bohr's efforts that Fermi was in the United States rather than his native Italy, and it was from Bohr that he learned that, theoretically, a bomb could be made. In 1938 anti-Semitic laws were passed in Italy. Because Fermi's wife was Jewish, they began looking for an opportunity to leave Italy. In the fall of that year, when Fermi was visiting Bohr's institute, Bohr told him that he would receive the Nobel Prize that year. Under ordinary circumstances this would have been a serious breach of etiquette; Nobel Prize winners are not supposed to be informed in advance. However, as Bohr realized, it would provide Fermi with a perfect opportunity. Fermi left Italy with his family in December 1938. After staying with Bohr for approximately two weeks, he took his wife and children with him to the Nobel ceremony in Stockholm and from there they sailed to New York, where a visiting professorship at Columbia University had been arranged.

Fermi was not the first physicist that Bohr helped in this manner. After Hitler came to power in 1933, Bohr made use of his contacts with the Rockefeller and other foundations to aid physicists who wanted to leave Germany. The money that was obtained was used to support them when they left Germany and came to Copenhagen. Once they were there, Bohr made use of his connections to get many of them academic posts in the United States and other countries. Before they got these posts they often spent a year or so doing scientific work at Bohr's institute.

In 1940 Germany invaded Denmark, and the country was quickly occupied when the small Danish army made only token resistance. Immediately after the occupation there wasn't a large German presence, and life remained more or less the same. Thus Bohr, whose mother was Jewish, did not feel the need to leave Denmark. However, in 1942 and 1943 there were an increasing number of anti-German strikes, demonstrations, and acts of sabotage, prompting the Germans to institute martial law in August 1943. The following month Hitler ordered that all the Jews in Denmark be deported, and on September 29 two German freighters arrived at Copenhagen. They were to transport the Jews to Germany.

However, on the previous day information about the plans for deportation had been given to Jewish leaders and the word quickly spread. By October 1 most of the Jews had gone into hiding in churches, in hospitals, and in private homes. When the German roundup began on October 1, only 284 Jews were captured. By comparison, 7,220 escaped to Sweden on private vessels such as fishing boats. Unexpectedly, help came from the German naval commander in charge of the Copenhagen port. He took his patrol vessels out of operation while the mass exodus was taking place, reporting to his superiors that the vessels were in need of repair.

HEISENBERG'S VISIT

In October 1941 there was a conference on astrophysics at the German Cultural Institute in Copenhagen. During the course of the meeting both Heisenberg, who was then director of the German atomic bomb project, and the German physicist Carl von Weizsäcker gave a public lecture. While he was in Copenhagen, Heisenberg came to Bohr's institute several times, and on one occasion Bohr and Heisenberg had a private conversation.

When the conversation ended, Bohr was very angry, and Heisenberg left shortly afterwards. At the time, neither man revealed what had been said and, as a result, the nature of the conversation between Bohr and Heisenberg became a matter of endless speculation. This speculation was fueled by the fact that, after World War II ended, Heisenberg claimed that he and other physicists had deliberately prevented the development of a German atomic bomb. On the other hand, Samuel Goudsmit, who headed the scientific team that investigated the progress that Heisenberg's project had made, believed that the reason little had been accomplished was that Heisenberg, a purely theoretical physicist who had no experience running large projects, had a poor understanding of the technical problems involved.

And then in February 2002 the Bohr family authorized the release of a letter that Bohr wrote to Heisenberg some years after the meeting, but never sent. The family had previously stipulated that Bohr's papers would be released 50 years after his death, in 2012. They released this letter 10 years early in order to quell some of the speculation. Bohr's letter was written in response to some claims made by Heisenberg that were quoted in Robert Jungk's 1958 book *Brighter Than a Thousand Suns*. "I think that I owe it to you to tell you that I am greatly amazed to see how much your memory has deceived you," Bohr wrote. Bohr then went on to say that he remembered his conversation with Heisenberg very well. He pointed out that Heisenberg had been convinced that Germany would win the war and had said that everything possible was being done to develop atomic weapons. Heisenberg, Bohr remarked, had also urged Danish cooperation with the Germans, the inevitable victors in the conflict.

Although Bohr was writing 17 years after the conversation took place, there are good reasons for thinking that his account was reasonably accurate. Heisenberg was probably not a good choice to be director of the German project. At the time the project began, all the basic physics concerning nuclear fission was known. Creating a bomb was more a matter of overcoming numerous practical difficulties than of discovering new physics. And when it came to practical matters, Heisenberg was often very inept. He had always concentrated on theory, so much so that when, as a young man, he defended his doctoral dissertation, one of the professors decided to ask him questions about basic laboratory physics. Heisenberg was asked, for example, how a battery worked, and he was unable to answer.

Once the bomb project was underway, Heisenberg developed a tendency to favor atomic reactor designs that looked good on paper but that turned out to be disasters in practice. He also seems to have grossly overestimated the quantity of fissionable material required to make a bomb.* During the course of the project Heisenberg's main scientific achievement was a theoretical one that had nothing to do with fission. He developed something called S matrix theory, which clarified certain problems regarding the atomic nucleus.

Bohr never made his account of the conversation public. Heisenberg had been a friend and colleague with whom he worked closely during the days that quantum mechanics was being developed, and he didn't want to fuel controversy. In the last paragraph of his unsent letter he assured Heisenberg that "this letter is essentially just between the two of us."

ESCAPE TO SWEDEN

In September 1943 the Swedish ambassador in Copenhagen told Bohr that he faced imminent arrest if he remained in Denmark. Bohr contacted friends in the Danish Resistance, who immediately arranged to have him and his family flee to Sweden. Because the boat could not accommodate all of them as well as others who were being ferried across, plans were made to take different members of the family across on succeeding nights. Thus some members of the Bohr family had to go into hiding for a day or two before they could leave.

---

*Heisenberg claimed that he deliberately made inflated estimates to impede development of the weapon. It is certainly possible that he had a change of heart and did this. However, he always claimed that he had tried to prevent development of the bomb from the beginning.

Bohr left Denmark around 10 P.M. on September 29 and arrived in Stockholm by train the next day. He was to go to England from there. On October 4 he left on a British Mosquito bomber, a plane that could avoid German fighter planes by flying at very high altitudes. But the plane had an engine malfunction and had to return. However, Bohr was able to leave on the following night.

During the flight, he sat on a pallet in the bomb bay. When an altitude of 30,000 feet was reached, the pilot told Bohr to put on the oxygen mask. But Bohr's flying helmet hadn't been properly adjusted, and he didn't hear the message. When the pilot received no reply from Bohr, he knew that something was wrong, so as soon as it was safe to do so, he descended to a lower altitude. When the plane landed, Bohr was in fine shape. He had slept most of the way, he said. Actually he had lost consciousness from lack of oxygen.

Shortly after arriving in England, Bohr was briefed on the progress that had been made toward developing an atomic bomb. At the time there were bomb projects in both the United Kingdom and the United States. The U.K. project was code named Tube Alloys, the U.S. one the Manhattan Project. Bohr was made a scientific adviser to Tube Alloys and an adviser to the American project also. He left London on November 28, bound for the United States, where he spent eight months at Los Alamos.

While he was associated with the Manhattan Project, Bohr traveled under the name Nicholas Baker and was called "Uncle Nick" by his colleagues. By this time he spoke English fluently; at the beginning of World War II, English had replaced German as the lingua franca at the institute. But Bohr had his own characteristic way of pronouncing the language, speaking for example of the "atomic bum."

He played only a small role in the development of the atomic bomb. "They did not need my help," he said later. However, he did contribute to the work that was being done, participating in some of the experiments and giving advice on theoretical matters. While Bohr was at Los Alamos, he began to think about the coming postwar period. As he did, he became convinced that the Soviets should be told about the bomb before it was used. Giving them the information would cause no harm; their physicists were working on a bomb project and they would have nuclear weapons soon enough anyway. But if the information were withheld, it would foster a climate of mistrust. On the other hand, if they were told, it would open the way for scientific cooperation in the future.

BOHR, ROOSEVELT, AND CHURCHILL

In February 1944, when he was in Washington, Bohr met Supreme Court Justice Felix Frankfurter, with whom he had had previous contacts regarding refugee problems. Upon learning that Frankfurter, who was an adviser to President Roosevelt, knew of the progress made by the Manhattan Project, Bohr outlined his ideas. Frankfurter replied that he believed Roosevelt might be responsive to them. In March, Frankfurter told Bohr that Roosevelt was also deeply concerned about the political impact of nuclear weapons and suggested that Bohr should go to London to present his views to Churchill. The appropriate arrangements were made, and Bohr arrived in London on April 14.

The meeting didn't go well. Churchill believed that the atomic bomb should be kept secret and he wasn't receptive to Bohr's arguments. Matters grew worse when the British physicist Lord Cherwell, who was also present at the meeting, made some remarks that Churchill took to be criticisms of the agreement that he and Roosevelt had made about British-American collaboration on the bomb project. Thus the conversation was sidetracked while the misunderstanding was cleared up. Bohr was unable to discuss all the matters that he had intended to bring up; too much of the meeting was spent discussing other issues. At the end of the meeting, Bohr asked to be allowed to write Churchill a letter elaborating on some points that he had wanted to make. Churchill replied, "It would be an honor to receive a letter from you," and then added, "but not about politics." Sometime later, in a note to Cherwell, Churchill wrote, "I did not like the man when you showed him to me, with his hair all over his head, at Downing Street."

After returning to the United States, Bohr had a meeting with Roosevelt, which seemed to go a lot better. Roosevelt seemed receptive to Bohr's ideas. However, when Roosevelt met privately with Churchill after the conclusion of an official meeting in Quebec, they decided that the bomb project should be kept secret. They also agreed that "enquiries should be made regarding the activities of Professor Bohr and steps be taken to ensure that he is responsible for no leakage of information, particularly to the Russians." Bohr had recently corresponded with a Russian physicist who had invited him to come to the Soviet Union. Bohr had declined, but the very fact that he had engaged in correspondence aroused suspicion. Neither Churchill nor Roosevelt was aware that the British physicist Klaus Fuchs had passed along information about the bomb to the Soviets. Not only did the Soviets know of the bomb project, they had enough details of the work being done at Los Alamos to reduce the time needed to bring their own atomic bomb project to a successful conclusion.

BACK TO DENMARK

Bohr returned to Copenhagen in 1945. He continued to do some research, but much of his time was now taken up by administrative work. The institute was being expanded, and new equipment was being installed. Plans had to be made concerning the lines of research that would be pursued at the expanded institute. He had other administrative duties as well. He had been president of the Royal Danish Academy of Sciences and Letters since 1939, and by now he was very much a celebrity, one of the best-known people in Denmark. The king and queen of Denmark, cabinet ministers, and foreign ambassadors all came to his home for dinner. And of course he also continued to have numerous visits by physicists.

In 1955 Bohr reached the mandatory retirement age for Danish university professors. However he had no desire to cease working, and he continued as director of the institute. By now much of his time was consumed accepting new honors. At the time of his death in 1962, he had been given more than 30 honorary doctorates and numerous medals. In 1947 the king of Denmark had awarded him the Order of the Elephant, which was ordinarily given only to heads of state.

On August 1, 1962, Bohr and his wife Margrethe celebrated their golden wedding anniversary. They embarked on a trip to Italy the following month, returning on October 27. On November 18 Bohr went upstairs after lunch to take a nap. Shortly afterwards he called out to Margrethe, who went upstairs and found him unconscious. A doctor was called, but Bohr was soon dead of heart failure. After his death Bohr was cremated, and his ashes were buried in the family grave. Shortly afterwards the Institute for Theoretical Physics was officially renamed the Niels Bohr Institute.

Bohr is considered one of the great physicists of the twentieth century, yet one of his most important contributions was to chemistry, not physics. He explained why the table of the elements exhibited a periodicity. And he explained why some of the elements (the rare earths) were exceptions. Furthermore, he solved the problem of why there were so many elements. The universe was not composed of some 90 basic building blocks, but of a small number of particles, which were the constituents of atoms.

The Continuing Search

At the beginning of the twentieth century, the quest for an understanding of the ultimate constituents of matter passed from the chemists to the physicists. By 1925 physicists thought they knew what the ultimate constituents of matter were and how they interacted. It was believed that there were just two particles: protons and electrons. Some of the electrons, it was thought, orbited the nucleus, while others existed *inside* the nucleus. By then it had been established that beta particles were electrons that were ejected from the nucleus, not from orbits around the nucleus, so that conclusion seemed natural. The theory also seemed capable of explaining how, for example, a helium nucleus could have a weight approximately equal to that of four protons, but a charge of only +2. The negative charge of two electrons cancelled out half of the positive charge of the four protons within the nucleus.

It was also believed that there were three basic forces in nature: gravity, electromagnetism, and a force that caused the particles within the nucleus to stick together. No one knew what kind of force the last one was. However, it had been established that electromagnetism couldn't possibly bind protons and electrons together, and the gravitational force between them was far too weak to be of any significance.

It was a simple picture of matter, as simple as the four-element theory with which chemistry had begun. And it was just as wrong. In fact, in 1925 physicists were already beginning to realize that there were serious problems with this model. For example, the electron was supposed to be approximately the same size as atomic nuclei. How could one confine numerous electrons in so small a space? Uranium, for example, was supposed to have 238 protons and 146 electrons within the nucleus.

And then there was an energy problem. Electrons confined within such small volumes would have to have implausibly large energies. Furthermore, electrons had spin. Therefore, nuclei should have spins of the same order of magnitude. But this was contradicted by experiment. At the time, nuclear spins had not yet been measured, but it was known that they were very small. There were other problems too, mostly of a highly technical nature, impossible to cover here without adding a whole book. However, I can sum matters up by saying that physicists were beginning to understand that if the proton-electron theory was correct, then the particles had to interact with one another in some very implausible ways.

THE NEUTRON

As the foremost nuclear physicist, Rutherford was very much aware of these problems. He realized that they might disappear if there were a third kind of particle, one that was electrically neutral, which he named the neutron. As early as 1920 Rutherford suggested that electrons might somehow combine with protons by a kind of nuclear alchemy, producing a new particle. In a 1920 lecture he remarked that this seemed "almost necessary to explain the building up of heavy elements."

Rutherford immediately put some of his Cavendish laboratory physicists to work looking for the hypothetical particle. One of them looked for neutrons in a tube filled with protons and electrons, through which an electrical discharge was passed. Meanwhile, another Cavendish physicist, James Chadwick, looked for the gamma rays that would theoretically be produced if the electron in a hydrogen atom fell into the nucleus. Everything that Rutherford and Chadwick could think of was tried, no matter how bizarre the idea might be. As Chadwick later remarked, some of the attempts were "so far-fetched as to belong to the days of alchemy."

But none of the experiments succeeded. Rutherford and Chadwick came close. They probably produced neutrons when they performed an experiment in which they bombarded aluminum with alpha particles. However, they saw nothing. In those days, neutral particles were hard to detect. Protons and electrons, which were electrically charged, could be manipulated with magnetic and electric fields. Neutrons couldn't.

Then, in 1932, Irène Joliot-Curie (the daughter of Marie Curie) and her husband, Jean Joliot, published a paper reporting that gamma rays were produced when paraffin was bombarded with alpha particles. When Rutherford and Chadwick read the paper, they didn't believe it. They suspected that what the two French physicists had seen was not gamma rays but neutrons.

Rutherford and Chadwick knew that if the Joliots realized that their conclusions were erroneous, they might discover the neutron first. So Chadwick immediately went to work performing new experiments. He soon found that if beryllium was bombarded with alpha particles, a kind of radiation consisting of particles with a mass close to that of the proton were produced. He ruled out the possibility that the radiation consisted of gamma rays by showing that, if it did, the gamma rays would have insufficient energy to produce the effects that were observed. Chadwick had discovered Rutherford's neutron.

Chadwick had beaten the Joliots to the discovery. Nevertheless, the story ended happily for all those involved. The Joliots continued to do important research with radioactive substances, for which they got the Nobel Prize for chemistry in 1935, the same year that Chadwick was given the Nobel Prize for physics.

MORE PARTICLES

In early 1932 the neutron wasn't the only new particle being talked about. The existence of two others had been suggested on theoretical grounds. These were the positron and the neutrino. Neither had been detected experimentally.

In 1925 Heisenberg had discovered a new theory called quantum mechanics, which described the behavior of subatomic particles much better than the old quantum theory. The following year the Austrian physicist Erwin Schrödinger had discovered another theory, which looked very different from Heisenberg's, but which turned out to be mathematically equivalent to it. But the new theory had one failing. Neither Heisenberg nor Schrödinger had been able to incorporate Einstein's special theory of relativity into quantum mechanics. This meant that Heisenberg's and Schrödinger's theory would be expected to give incorrect results when particles were moving very rapidly. There are two theories of relativity. The first, which Einstein discovered in 1905, is called the special theory of relativity. It deals with the behavior of bodies that are moving very rapidly, at velocities that are a significant fraction of the speed of light. The second, called the general theory of relativity, is Einstein's theory of gravitation.

The problem of combining quantum mechanics and special relativity was solved by the English physicist P.A.M.* Dirac in 1928. Dirac

---

*Not even Dirac's closest colleagues knew that the initials stood for "Paul Adrian Maurice." Naturally they tried to guess his name, but no one ever succeeded. Once a colleague suggested that the M might stand for "Mussolini." Of course this was a joke, probably made in frustration at not being able to get any clues about what the name really was.

was one of the great theoretical physicists of the twentieth century but he got into the field by accident. As an undergraduate at Bristol University he studied engineering, intending to get a job in the field when he graduated. But England was in recession when he got his degree, and Dirac was unable to find employment so he decided to pursue graduate studies in mathematics, a subject in which he had always had a strong interest.

There were two courses of study in Bristol's graduate program, called pure mathematics and applied mathematics. Dirac was indifferent as to which one he would study. Now it happened that there were only two students in Bristol's advanced mathematics program at the time and, when the other student chose applied mathematics, Dirac did too. He didn't want to make the professors teach two different sets of courses.

What was called "applied mathematics" in England was really theoretical physics. At the time only experimental physics was taught in physics departments. Traces of this system remain to this day. For example, the British physicist Stephen Hawking is Lucasian Professor of Mathematics at Cambridge University, not a professor of physics.

When Dirac completed work on his theory in 1928, it was a notable success. Among other things, it explained electron spin, which turned out to be a relativistic effect, rather than something analogous to the spin of a macroscopic object like a top. But the theory also made what seemed to be a very strange prediction. If Dirac's theory was correct, then there had to exist particles that had properties like the electron, but that carried a positive rather than a negative charge. At the time, such particles, called positrons, had never been observed.

COSMIC RAYS

The American physicist Carl Anderson received his Ph.D. from the California Institute of Technology in 1930, after which he immediately set to work studying cosmic rays. Cosmic rays are not a kind of radiation like gamma rays. They are rapidly moving electrons and atomic nuclei that fall on the Earth from space. In the 1930s they provided physicists with a kind of high-energy "laboratory" for doing research on subatomic particles. At the time, nothing resembling today's particle accelerators existed. Laboratory devices that could accelerate particles could impart only relatively low energies.

Before Anderson began his experiments he constructed a cloud chamber (a device that allowed physicists to see the paths that particles followed) with the most intense magnetic field that had ever been achieved. Magnetic fields cause charged particles to follow curved paths. By studying the paths, physicists can draw conclusions about their mass.

In 1932, the year that the neutron was discovered, Anderson detected particles that behaved as though they were electrons, except that their paths curved in the wrong direction. This meant that they must carry positive charge. They clearly couldn't be protons. Protons had masses far larger than the particles he was seeing. So he suggested that it was a new particle, a kind of electron with a positive charge. At the suggestion of the editor of the journal to which he submitted his paper, he called it a "positron." It would be more accurate to use the name "antielectron" instead. Today we know that the positron is the antiparticle of the electron. When the two collide, they annihilate one another and a small burst of gamma rays appear in their place. However, by the time that a theory of antiparticles had been developed, the term "positron" had been in use for so long that physicists were reluctant to change it.

THE NEUTRINO

When physicists began to study the alpha and beta particles that were ejected from atomic nuclei, they encountered a puzzle. The alpha particles that were produced by any given radioactive substance all had the same energy; they flew off at the same velocity. But this was not the case with beta particles. Some were faster moving and more energetic than others. This seemed paradoxical because the same amount of energy should have been released in any radioactive disintegration of a given element. Physicists suggested a variety of different explanations for this phenomenon. Bohr went so far as to suggest that the hallowed law of conservation of energy might sometimes fail on the subatomic scale. Most physicists were unwilling to accept that. However, the other ideas that were proposed didn't seem any better.

Then in 1930 Wolfgang Pauli suggested a "desperate way out." The problem would disappear, he said, if one assumed that a new, as-yet-undetected particle was emitted at the same time as the electron (beta particle). The reason that the new particle remained undetected, Pauli went on, was that it had about 10 times the penetrating power of gamma rays. Pauli knew that his idea was very speculative, and he had no desire to rush into print. Instead, he outlined his hypothesis in a letter sent to a radioactivity conference that he could not attend. Soon he began to discuss the matter with other physicists, while wavering between believing that his hypothesis might be correct and thinking that it was assuredly wrong.

Pauli called his hypothetical particle the "neutron." Naturally it bore no relationship to the "neutron" of which Rutherford spoke, which Chadwick was soon to discover. But the confusion that might have resulted was avoided when Enrico Fermi suggested adding the Italian diminutive suffix "-ino" to the name. Pauli's particle was thus christened the neutrino ("little neutral one").

In physics, speculative ideas often begin to seem much more reasonable when there is a theory to explain them. In 1933 Fermi advanced just such a theory, proposing that the electron and the neutrino were spontaneously created at the moment that a radioactive disintegration took place. At the same time, one of the neutrons in the nucleus was changed into a proton. Fermi showed that electron-neutrino production could be explained if one assumed the existence of a new force (now called the weak force). He concluded that the mass of the neutrino was probably near zero. This would explain why it hadn't been detected.

It sounds as though Fermi's ideas were even more audacious than

Pauli's, and indeed many scientists thought so. Fermi's first paper on the subject was rejected "because it contained speculations too remote from reality to be of interest to the reader." Yet Fermi's theory rapidly gained acceptance. It was based on widely accepted theoretical principles, and it was developed in a clear and logical way. The theory explained how a nucleus could eject an electron even though it was composed only of protons and neutrons. And finally, even though no neutrinos were seen in experiments, Fermi's theory was consistent with experimental data.

The concept of the neutrino gained acceptance long before the particle was ever observed. There was no experimental evidence for its existence, but particle physics couldn't do without it and in the end Pauli and Fermi were vindicated. The neutrino was finally discovered in 1953, the year before Fermi's death and five years before Pauli's.

MESONS

Quantum mechanics tells us that the forces of nature can manifest themselves as particles. Photons of light, for example, are the particles associated with the electromagnetic force. Such particles are not constituents of matter; they have an entirely different character. The photons that are responsible for electromagnetic forces are never observed. They are emitted by one matter particle and absorbed by another before they can be detected. However, there is little doubt that these phenomena take place because there is ample indirect evidence that they do.

In 1934 the Japanese physicist Hideki Yukawa postulated the existence of yet another force particle, which he called the meson. In 1932 Yukawa began his academic career with an appointment at Osaka Imperial University, which had been founded the previous year. The discovery of the neutron and the publication of Fermi's theory started him thinking about the nature of the force that bound protons and neutrons together in an atomic nucleus. He realized that, though the exact nature of this force was unknown, one could deduce things about it. For example, it had to be a force that acted only over very short distances. No experiment had ever suggested that it had any effects outside the nucleus.

Yukawa proceeded by writing down a mathematical formula for the force. It wasn't especially difficult to do this. He looked for the simplest mathematical form that was consistent with experimental facts. He knew that, if necessary, refinements could be added later. Then, applying the principles of quantum mechanics, he deduced that, if the force did have that form, there had to exist a previously unobserved particle that had a mass approximately 200 times greater than that of the electron.

Yukawa published a paper on his theory in 1934. Three years later Anderson discovered a new particle in cosmic rays. Its mass was in line with Yukawa's predictions, and at first physicists believed that Yukawa's theory had been confirmed. In reality, it hadn't been. It turned out that the new particle was *not* Yukawa's meson. It had about the right mass, but its other properties were inconsistent with his theory.

Anderson's particle, which is now called the muon after the Greek letter mu, was actually a kind of "heavy electron." It had a large mass, but it otherwise exhibited electronlike properties. This was a great puzzle to the physicists of the day, because there seemed to be no reason it should exist. It was not a component of ordinary matter. It could be observed in the high-energy cosmic ray "laboratory," but it quickly decayed (that is, disintegrated) into other particles.

The English physicist Cecil Powell discovered Yukawa's meson in 1947. Powell found evidence of its existence in photographic plates that had been exposed to cosmic rays in the Bolivian Andes. The particle was found to be a little heavier than the muon, and it interacted strongly with nuclei, as Yukawa's particle was expected to do. Unlike the muon, which always carried a negative charge, the new particle could have either a positive or a negative charge, or it could be electrically neutral.

At this point it is necessary to digress a little about nomenclature. When the muon was thought to be Yukawa's particle, it was called the "mu meson." When the term "meson" was later reserved for particles that interacted strongly with nuclei, "meson" was dropped from the name, and it became the muon. When Powell discovered Yukawa's particle, he called it the "pi meson" to distinguish it from the particle discovered by Anderson. This name was later shortened to "pion." The pion, however, is still classed among the mesons, while the muon isn't.

THE EIGHTFOLD WAY

In 1925 physicists had known of two particles of matter, the proton and the electron. By 1947 five were known: the proton, neutron, electron, muon, and neutrino. Yet another particle, the pion, was known, but it was associated with a force and it wasn't a component of matter. In 1925 they had known of two forces: gravity and electromagnetism. Now there were four. The two new ones were the weak nuclear force (now called simply the "weak force") and a strong nuclear force that held nuclei together.

Matters were still relatively simple, but they weren't to remain that way for long. Beginning in 1948, physicists began to discover numerous other new particles. The development of the cyclotron—and later more advanced particle accelerators—meant that it was now possible to produce the energies needed to study particle physics in the laboratory. Scientists were no longer dependent upon cosmic rays.

By the end of the 1950s, hundreds of new particles were known. It was found that there were numerous kinds of baryons (heavy particles with properties similar to those of protons and neutrons) and a wide variety of mesons as well. All of the new particles were unstable. They decayed into lighter particles in tiny fractions of a second. Nevertheless, it was impossible to deny their existence. Matters seemed far worse than when the chemists were confronted with the existence of 50 or 60 elements. It seemed absurd to consider all of the new particles to be "elementary." Yet as far as anyone knew, they had no smaller components.

The person who did the most to bring order to this chaos was the American physicist Murray Gell-Mann. Gell-Mann was born in New York in 1929, the son of an Austrian immigrant. A precocious youngster, he began teaching himself calculus at the age of seven, and he entered Yale University when he was 15. Like Dirac, Gell-Mann seems to have entered the field of physics by accident. He wanted to study archeology in college, but his father wanted him to enter a field like engineering, where the prospects of employment were better. The two finally found a compromise: Gell-Mann studied physics and earned a Ph.D. in physics from the Massachusetts Institute of Technology in 1951.

Gell-Mann began his career in physics when more and more new particles were being discovered. In the early part of his career, he made some important contributions to the field of particle physics. But it was only somewhat later that he began to ponder the problem of making some sense of the multitude of particles that were then known.

It wasn't yet possible to explain the existence of so many particles. But in 1961 Gell-Mann and the Israeli physicist Yuval Ne'eman independently showed that it was possible to discern order in the tables of particles, just as Mendeleev had found order in the table of the elements. The scheme, which Gell-Mann called the eightfold way, grouped mesons and baryons into families. He called the organizing scheme eightfold because the most commonly observed mesons and baryons were grouped in sets of eight particles each. Of course the name was also a pun. The original eightfold way was a recipe for reaching enlightenment that was devised by the Buddha in the fifth or sixth century B.C.

The eightfold-way theory had its first success when Gell-Mann used it to predict the existence of an as-yet-undiscovered particle, just as Mendeleev had predicted the existence of undiscovered evidence. Gell-Mann called the hypothetical particle the omega-minus (it had a negative charge). In 1964 experimentalists found the particle and confirmed that it had exactly the mass Gell-Mann said it would have.

However, particle physics had not reached nirvana yet. The next task was to discover why the eightfold scheme worked. The explanation called for something like what Bohr had done when he found a way to explain the periodicities in the table of the elements. A solution to this problem was found in 1964 when Gell-Mann and the American physicist George Zweig independently discovered that the eightfold-way theory could be explained if it was assumed that mesons and baryons were made up of even smaller particles, which Gell-Mann called quarks and which Zweig called aces. According to the two physicists, mesons were made up of two quarks (or aces), while each baryon had three constituents. The term "quark," which Gell-Mann had taken from a passage in James Joyce's novel *Finnegan's Wake*, quickly won out over its competitor and "ace" fell into disuse.

Initially it was thought that there were just three kinds of quarks, which were given the names "up," "down," and "strange." The names should not be taken literally. They were just labels. If the three quarks were called Tom, Dick, and Harry, nothing would change. Perhaps "strange" deserves a few words of explanation. Before the quark theory had ever been thought of, Gell-Mann had used the whimsical term "strangeness" to describe a mathematical property possessed by certain particles. There was nothing especially strange about them except that they didn't disintegrate as quickly as other particles. Today it is known that there are actually six kinds of quarks. The other three are called "charm," "bottom," and "top." For a while there was a movement to call the last two "truth" and "beauty." However, the more prosaic names won out. Protons, neutrons, and the more common mesons are made of up and down quarks. The remaining four are found only in particles that are created in the laboratory.

THE HUNTING OF THE QUARK

When Gell-Mann and Zweig proposed their theories, quarks had never been observed in nature, so many physicists were skeptical of their existence. Their skepticism deepened when extensive experimental searches failed to turn up any evidence of free quarks in nature. The quark theory seemed to work, but the doubters viewed them as little more than mathematical fictions, saying that particles behaved *as if* they were made up of quarks, just as some nineteenth-century skeptical chemists had taken the view that matter behaved as if it was made up of atoms.

The experiments went on, however, and in 1968 experiments at the Stanford Linear Accelerator Laboratory showed that quarks were indeed real. When protons were bombarded with high-energy electrons, pointlike charges were discovered inside the proton. These charges could only be charged particles, in other words, quarks.

There were still problems, however. Physicists had never succeeded in gaining a good theoretical understanding of the so-called strong force, the force that held protons and neutrons together in nuclei. They had devised various approximations that described this force but none was entirely accurate. It was now apparent why they had failed. The strong force was actually the result of forces between quarks *inside* protons and neutrons. No one yet knew what these forces were, but there was every reason to think that their nature would sooner or later be discovered.

The first step toward understanding these forces was made in 1964 when University of Maryland physicist Oscar W. Greenberg suggested that certain theoretical problems concerning quarks could be solved if it was assumed that quarks possessed a quality called "color" that was analogous to electrical charge. But there was an important difference between electrical charge, which can only be positive or negative, and the color charges on quarks. The color charges came in three varieties, not two. The three color charges were duly named red, green, and blue, after the three primary colors of light. Those names should not, of course, be taken literally. Quarks have no color in the everyday sense of the term. Color is a property of light, which is emitted by atoms. Quarks, which are constituents of constituents of atoms, do not emit, absorb, or reflect light.

During the mid-1970s, theoretical physicists worked out the color theory in detail, gaining an understanding of the forces binding quarks together. This theory, which was called quantum chromodynamics, also explained why free quarks were not seen. While the other known forces decreased with increasing distance, the forces between quarks grew larger as they moved apart, effectively confining them within mesons and baryons.

LEPTONS

Particles such as electrons and muons are not made up of quarks, and they are thus insensitive to the strong force. It is electrical attraction between unlike charges that binds electrons in atoms. Both electrons and muons belong to a class of particles called leptons, and there are six of them, just as there are six quarks. The six leptons are the electron, the muon, the tauon (named after the Greek letter tau), and three different kinds of neutrinos, which are called electron neutrino, muon neutrino, and tauon neutrino.

Like the muon, the tauon has electronlike properties. However it is much heavier than its two counterparts. It weighs about 3,500 times as much as the electron and about 170 times as much as the muon. The neutrinos, on the other hand, are very light, so light that, as I write this, their mass has not yet been determined. It wasn't until 1998 that it was even established that neutrinos *had* any mass.

Thus there are 12 matter particles, the six quarks and the six leptons (antiparticles such as the positron are not counted separately). Although this is a vast improvement over the situation in 1960, when physicists had hundreds of particles to contend with, scientists do not yet believe that they have found the key to the universe. Simply too much remains to be explained. Physicists don't know why there should be 12 matter particles, and not more or fewer. They don't know why the particles have the masses they do. And they don't know why there should be four forces in nature, rather than three or six, or why they have the strengths that they do.

The four forces are gravity, electromagnetism (electrical attraction and repulsion and magnetic forces are explained by the same theory), and the strong and weak nuclear forces. The weak force is responsible for certain kinds of nuclear reactions, including beta decay; the name reflects the fact that it is much weaker than its counterpart, the strong force. There is an enormous disparity between the strength of the forces. For example, the strong force between two protons in a nucleus is about 10^{39} times stronger than the gravitational force between them.* No one has any good idea why the difference should be so great.

Physicists would like to have some theory that explains the properties of the particles and of the four forces. As I write this, no such theory exists. But there is hope that one will eventually be found.

SUPERSTRINGS AND M THEORY

Physicists classify elementary particles by the ways they respond to different forces. Recall that quarks are subject to the strong force, while leptons are not. This seems to suggest that if physics found a theory that gave a unified explanation of all four forces, much would be learned that was new. In particular, physicists might gain an understanding of why particles have the properties they do.

There are precedents that indicate this is indeed likely to be the case. For example, around the middle of the nineteenth century, the Scottish physicist James Clerk Maxwell developed a theory that gave a unified explanation of the phenomena of electricity and magnetism. Not only did the theory suggest that light was made up of electromagnetic waves, it also led to the discovery of such new phenomena as radio transmission.

Similarly, when, in 1967, the American physicist Steven Weinberg and the Pakistani physicist Abdus Salam independently discovered a theory that unified the weak and electromagnetic forces, they found that the theory predicted the existence of certain undiscovered particles (these were particles that transmitted forces, not matter

---

*10^{39} is the number represented by the numeral "1" followed by 39 zeros.

particles). These particles were found in experiments performed in 1983. If the unification of just two forces can produce such striking results, the gains would most likely be spectacular if a theory were found that unified all four forces. The problem is very difficult, however, and none of the various approaches has yet led to success, although some progress has been made.

As I write this, the most promising line of attack seems to be superstring theory. Superstring theory is a mathematically very difficult theory that postulates the existence of extra dimensions of space. In the original version of the theory it is assumed that there are 10 dimensions in all: nine spatial dimensions and one of time. The extra dimensions are believed to be compacted, or rolled up, to dimensions far smaller than those of an atomic nucleus. In order to visualize this, imagine making a cylinder of a sheet of paper and then rolling up the paper ever more tightly. As this is done, the diameter of the cylinder progressively decreases. It the paper could be rolled up enough, it would resemble not a two-dimensional sheet but rather a very thin rod.

What causes the extra dimensions to be rolled up? The physicists who work in superstring theory think this is the wrong question to ask. They suspect that *all* of the dimensions were originally compacted and that the familiar three spatial dimensions we know became uncompacted early in the history of the universe. Indeed, there are some theories that explain how this could have happened. The theory postulates tiny vibrating loops called superstrings and posits that the various particles of matter are just superstrings vibrating at different frequencies. In other words, there are not 12 elementary particles of matter; there is just one fundamental kind of entity. Superstrings have not been observed, and they might never be seen in the laboratory. If they do indeed exist, they have dimensions far smaller than an atomic nucleus, making it impossible to detect them with any conceivable kind of scientific apparatus. They could presumably be seen with a particle accelerator that was powerful enough. But such an accelerator would have to be larger than our galaxy.

When I said earlier that superstring theory was "mathematically very difficult," I was understating the problem. It is so difficult that, as I write this, no one yet knows what the mathematical equations of the theory are. Only approximate equations are known, and these are so complicated that they can be solved only approximately. The extra dimensions add complications too, because these dimensions can be interwoven in different ways. And because there is not one superstring theory but five, thousands of different versions of the theory are possible.

It might appear that the task of using superstring theory to describe the four forces and to explain the properties of the 12 particles of matter is hopeless. In reality, it might not be. In 1995, at a conference on superstrings held at the University of Southern California, Princeton University physicist Edward Witten announced that he had made a new discovery. He had found that the five superstring theories were closely related to one another and to something called supergravity theory as well. The six theories simply described the same physics in different ways. Furthermore the theories were all related to an 11-dimensional theory called M theory.* There was one catch. Witten didn't know what M theory was. And as I write this, no one else does either. Witten showed that M theory had to exist, but he was unable to deduce the mathematical principles on which it was based. However, his discovery did give a boost to superstring research. The relatedness of the six theories meant that, if a problem was too difficult to solve using a particular theory, one of the others could always be used in its place.

However the physicists who do research on superstring theory still face numerous problems. For example, they can't even be sure whether superstrings are entities that exist in space and time, or whether space and time might be made of superstrings. Some progress has been made in determining the ways in which the extra

---

*Witten didn't specify what the "M" stood for.

dimensions might interweave with one another but much remains to be done. And of course there is always the problem that superstring theories are mathematically the most difficult theories ever discovered.

DECIPHERING THE UNIVERSE

Since the time of Paracelsus, scientists of almost every generation have sought the key to the workings of universe. Paracelsus thought he had found it in the principles of sulfur, mercury, and salt, while others clung to the theory that a complete explanation of the makeup of the world was provided by the four-element theory. Robert Boyle attacked both theories in *The Sceptical Chymist* and introduced the ideas out of which the modern theory of the chemical elements grew. Boyle was succeeded by a series of brilliant chemists who discovered new elements, gained new insights into chemical reactions, and finally disproved the four-element theory once and for all.

Mendeleev, the chemical magician whom the English called "Faust," found order in the table of the elements and predicted the properties of elements that were yet to be discovered. In the next century, Bohr was able to explain why periodicities in the properties of the elements existed.

By then the quest had passed from the hands of the chemists into those of the physicists, who made a series of discoveries that were eerily like those that had been made in chemistry. Physicists discovered new particles until the number of known particles grew beyond reason. Then, like their predecessor Mendeleev, Gell-Mann and Ne'eman discovered a hidden order. Like Bohr, who had probed the workings of the atom, Gell-Mann and Zweig theorized about the inner mechanisms of mesons and baryons, introducing the concept of the quark.

In our own time the quest has been taken to an even deeper level, notably by physicists working in superstring theory. As I write this, no one can say whether the superstring theorists will ultimately be successful. Perhaps some other way of unifying the four forces will be found. Alternatively the superstring theorists might eventually discover what M theory is and solve some or all of the problems currently facing physics. But even if they do, there is no guarantee that the quest will have come to an end. During the early days of superstring research, many physicists in the field proclaimed that a successful theory would be a "theory of everything," that is, a theory that would explain all the known laws of physics. But nowadays superstring theorists generally make a more sober assessment, admitting that they don't know whether it would be a final theory.

It might turn out that no ultimate theory will ever be discovered, that scientists will find themselves probing the workings of nature at ever deeper levels. After all this is what has always happened before. Scientists originally thought that the chemical elements were the fundamental components of the universe. Then the physicists showed that they were not, by discovering first that atoms were composed of smaller particles and then that some of these particles themselves had components. Then superstring theory introduced the idea that even smaller entities were fundamental.

The search for the key to the universe might eventually come to an end. But it is also possible that understanding the workings of nature is like peeling off the layers of an onion one by one and that the quest will go on for as long as there are scientists.

APPENDIX

A Catalog of
the Elements

Among the known elements, 91 occur naturally. The remainder represents an increasing number of new and artificially produced elements that reflect the innovative work of modern scientists. Arranged by atomic number and with a brief description, here is a comprehensive list of the known elements.

Element	Symbol	Atomic No.	Year isolated
Hydrogen	H	1	1766
Helium	He	2	1868
Lithium	Li	3	1817
Beryllium	Be	4	1798
Boron	B	5	1808
Carbon	C	6	
Nitrogen	N	7	1772

Discoverer	Comment
British chemist Henry Cavendish	The most abundant element; formed concurrent with the universe; used in balloons, airships, and by deep-sea divers.
French astronomer Pierre Janssen	First observed on the sun, this inert and noble gas forms no known stable compounds.
Swedish chemist Johan August Arfvedson	Lightweight, reactive alkali metal whose isotope lithium-6 was used to build the hydrogen bomb.
French chemist Louis-Nicolas Vauquelin	Rare alkaline-earth metal derived from the mineral beryl of which emeralds and aquamarines are crystals.
British chemist Sir Humphry Davy and French chemists Joseph-Louis Gay-Lussac, Louis-Jacques Thenard	Nearly as hard as diamond, this brittle crystal is rare in pure form; combines to form borax; also valuable in the production of glass and semiconductors.
Known to the ancients	Exists in multiple forms, including diamond and graphite; found in nearly all of the compounds that constitute living things.
British chemist Daniel Rutherford	In pure form, a relatively unreactive gas; makes up about three-quarters of the Earth's atmosphere; important constituent of deoxyribonucleic acid (DNA) and proteins.

Element	Symbol	Atomic No.	Year isolated
Oxygen	O	8	1774
Fluorine	F	9	1886
Neon	Ne	10	1898
Sodium	Na	11	1807
Magnesium	Mg	12	1808
Aluminum	Al	13	1827
Silicon	Si	14	1824

Discoverer	Comment
British chemist Joseph Priestley	Among the most abundant and most important elements for life; reacts easily with most other elements, including hydrogen to form water.
French chemist Henri Moissan	Highly reactive; used in the manufacture of Teflon and in toothpaste.
British chemist Sir William Ramsay	Colorless gas that forms no known compounds; glows when electrified and used in commercial signage.
British chemist Sir Humphry Davy	Highly reactive metal, easily liquefied and valuable in transferring heat from nuclear reactors; forms many important compounds, including table salt.
British chemist Sir Humphry Davy	Lightweight metal with anticorrosive properties; combines with iron and aluminum to form metal alloys useful in manufacturing.
Danish chemist Hans Christian Oersted	Found in the ore bauxite; is the Earth's most abundant metal; industrial use includes many items, from airplanes to soda cans.
Swedish chemist Jöns Jakob Berzelius	Combined with oxygen, it forms an important component of sand, rock, and soil; appears also as quartz; used to make glass, semiconductors, and ceramics.

Element	Symbol	Atomic No.	Year isolated
Phosphorus	P	15	1669
Sulfur	S	16	
Chlorine	Cl	17	1774
Argon	Ar	18	1894
Potassium	K	19	1807
Calcium	Ca	20	1808
Scandium	Sc	21	1879

Discoverer	Comment
German physician Hennig Brand	First used as a light source, its name comes from the Greek meaning "bearer of light"; one form, "white phosphorus," is highly explosive; used in the manufacture of fertilizers.
Known to the ancients	Brittle, yellow, reactive nonmetal used mainly to produce sulfur dioxide; useful in textile manufacturing, paper making, and as a food preservative.
Swedish chemist Carl Wilhelm Scheele	Poisonous, diatomic gas occurring in salt water and salt mines; useful as a disinfectant and in the manufacture of industrial plastics such as PVC.
British physicist Lord Rayleigh and Scottish chemist William Ramsay	Noble, inert gas that forms no known compounds; used to fill the cavity of light bulbs, as its unreactive quality prevents filament corrosion.
British chemist Sir Humphry Davy	Reactive, alkali metal; used to make fertilizers and as potassium nitrate, or saltpeter, explosives.
British chemist Sir Humphry Davy	Reactive metal that easily forms many compounds found in limestone, lime, marble, and gypsum; important component of bones and coral reefs.
Swedish chemist Lars Fredrik Nilson	Lightweight, anticorrosive metal that forms few compounds; one of its isotopes is useful as a tracer of specific constituents in oil refining.

Element	Symbol	Atomic No.	Year isolated
Titanium	Ti	22	1791
Vanadium	V	23	1801
Chromium	Cr	24	1797
Manganese	Mn	25	1774
Iron	Fe	26	
Cobalt	Co	27	1739
Nickel	Ni	28	1751

Discoverer	Comment
British mineralogist William Gregor	Dense, lightweight, anticorrosive metal stronger than steel; used in aircraft engines; titanium dioxide used widely in paints and plastics.
Mexican mineralogist Andres Manuel del Rio	Shiny, anticorrosive metal; some of its compounds are used to manufacture sulfuric acid and a harder form of steel known as vanadium steel.
French chemist Louis-Nicolas Vauquelin	Hard metal that is white in pure state but brilliantly colored as compounds; in trace amounts it gives gems their colors; used in paints, automobile bumpers, and stainless steel.
Swedish chemist Carl Wilhelm Scheele	Hard, ironlike metal; used to add hardness to steel and in battery manufacture.
Known to the ancients	Metal seldom found in pure state and often combined with oxygen; contained chiefly in hematite and magnetite; essential to steel making.
Swedish chemist Georg Brandt	Rare metal easily magnetized and used in the alloy alnico to make industrial magnets; its isotope cobalt-60 produces radiation that is a treatment for cancer.
Swedish chemist Axel Fredrik Cronstedt	Along with iron, is believed to be one of the two main elements that make up the Earth's core; hard and anticorrosive, it forms several important industrial alloys.

Element	Symbol	Atomic No.	Year isolated
Copper	Cu	29	
Zinc	Zn	30	1746
Gallium	Ga	31	1875
Germanium	Ge	32	1886
Arsenic	As	33	1250
Selenium	Se	34	1817

Discoverer	Comment
Known to the ancients	Soft metal; excellent conductor of electricity; name derives from the Latin *cuprum*, meaning "from Cyprus," where it was mined by the Romans; alloys are brass and bronze.
German chemist Andreas Marggraf	Hard, brittle metal; prevents corrosion in steel by a process known as galvanization; used also to make dry-cell batteries and electronic devices.
French chemist Paul-Émile Lecoq de Boisbaudran	Soft metal similar to aluminum; gallium arsenide is used extensively in laser lights, electronic displays, CD players; used to detect subatomic particles known as neutrinos.
German chemist Clemens Winkler	A metalloid chemically similar to silicon; in a process called doping, certain impurities are added that make it useful in the manufacture of transistors.
German, Albertus Magnus	Brittle metalloid found in mineral compounds realgar and orpiment; although poisonous it has been developed into several medicinal compounds; essential to the electronics industry.
Swedish chemist Jöns Jakob Berzelius	Metalloid that exists as two allotropes, one glasslike and the other metallike; photoconducting properties useful in photocopy machines; included in the diet, it has health benefits.

Element	Symbol	Atomic No.	Year isolated
Bromine	Br	35	1826
Krypton	Kr	36	1898
Rubidium	Rb	37	1861
Strontium	Sr	38	1789
Yttrium	Y	39	1789
Zirconium	Zr	40	1787
Niobium	Nb	41	1801

Discoverer	Comment
French chemist Antoine-Jerome Balard	Along with mercury, one of two elements that at room temperature is found as a liquid; some compounds are used as pesticides and to make photographic film.
British chemist Sir William Ramsay	Noble gas that gives off a bluish light when subjected to electrical charge; often used for airport runway lighting; forms only one known compound, krypton fluoride.
German chemists Robert Bunsen and Gustav Kirchhoff	Soft metal so reactive that it burns when exposed to air; of little commercial value.
British scientist Adair Crawford	Metal whose isotope, strontium-90, is the by-product of nuclear explosions; as a compound, it is added to fireworks and flares to produce red color.
Swedish scientist Johan Gadolin	Active metal rare on Earth, but plentiful in rocks brought back from the moon; its compounds are used in laser technology and in high-temperature superconductors.
German chemist Martin Heinrich Klaproth	Durable metal highly resistant to heat; its mineral zircon, a compound of zirconium silicate, is a semiprecious gem often substituted for diamonds.
British mineralogist Charles Hatchett	Soft metal found in the mineral columbite, along with iron and manganese; combined with germanium it forms an excellent high-temperature superconductor.

Element	Symbol	Atomic No.	Year isolated
Molybdenum	Mo	42	1778
Technetium	Tc	43	1937
Ruthenium	Ru	44	1844
Rhodium	Rh	45	1803
Palladium	Pd	46	1803
Silver	Ag	47	

Discoverer	Comment
Swedish chemist Carl Wilhelm Scheele	Hard silvery metal mined from the ore molybdenite; added to steel as "moly steel" to increase ability to withstand pressure and temperature shifts in auto and plane engines.
Italian scientists Emilio Segrè and Carlo Perrier	Discovered when molybdenum was bombarded with the nuclei of a hydrogen isotope; the first man-made element, hence its name, which means "artificial" in Greek.
Estonian scientist Karl Karlovitch Klaus	The by-product of platinum refining; useful as a catalyst in many industrial processes; as an alloy, used for fountain pen nibs and electrical contacts.
British scientist William Hyde Wollaston	Hard, anticorrosive metal whose salts have a red color, its name is derived from the Greek word for "red"; used in automotive catalytic converters.
British scientist William Hyde Wollaston	Anticorrosive, soft metal often found combined with platinum; useful in dentistry, as a cancer-fighting agent; easily absorbs hydrogen and used as a purifier of that gas.
Known to the ancients	Soft, malleable metal; excellent conductor of heat and electricity; long used as currency; made into jewelry and eating utensils; used in photographic film.

Element	Symbol	Atomic No.	Year isolated
Cadmium	Cd	48	1817
Indium	In	49	1863
Tin	Sn	50	
Antimony	Sb	51	
Tellurium	Te	52	1782
Iodine	I	53	1811

Discoverer	Comment
German pharmacist Friedrich Strohmeyer	Soft metal chemically similar to zinc; used in electroplating and for rechargeable batteries; neutron-absorbing properties make it useful in nuclear power plants.
German chemist Ferdinand Reich	Soft bluish metal and by-product of zinc refining; used chiefly as an alloy in the manufacture of transistors and as an indicator of chemical activity in nuclear reactors.
Known to the ancients	Soft metal mined from the ore cassiterite; highly malleable and useful as an alloy to form pewter and bronze and as tin plate to protect steel from corroding.
Unknown	Hard, brittle metal; poor conductor of electricity; used chiefly in the production of safety matches and as a flame retardant for the plastic PVC.
German mining inspector Franz Joseph von Reichenstein	Brittle metalloid; poor conductor of electricity; combines with gold to form telluride; combined with other metals, it makes them easier to machine.
French chemist Bernard Courtois	Highly reactive solid that vaporizes to a violet-colored gas; found in seaweed; used as an antiseptic, in salt compounds as a dietary supplement, and in the production of photographic film.

Element	Symbol	Atomic No.	Year isolated
Xenon	Xe	54	1898
Cesium	Cs	55	1860
Barium	Ba	56	1808
Lanthanum	La	57	1839
Cerium	Ce	58	1803
Praseodymium	Pr	59	1885
Neodymium	Nd	60	1885

Discoverer	Comment
British chemist Sir William Ramsay	Noble gas that, unlike most, forms several compounds; used like neon to produce tube lighting.
German chemists Gustav Kirchhoff and Robert Bunsen	Softest metal known; some compounds highly reactive with water and carbon dioxide releasing oxygen and making it useful in breathing apparatuses for firefighters.
British chemist Sir Humphry Davy	Soft, reactive, abundant metal; as barium sulfate, it blocks transmission during diagnostic X rays to highlight organs and other tissue.
Swedish chemist Carl Gustaf Mosander	Silvery, reactive, malleable; first of the rare earth elements, whose added electrons are hidden in their atoms' interiors; as a compound used in high-intensity lighting.
Swedish chemist Jöns Jackob Berzelius, German chemists Wilhelm von Hisinger and Martin Klaproth	Most abundant of the rare earths; named for an asteroid; compounds and oxides used in lighting, self-cleaning ovens, cameras, telescopes.
Austrian mineralogist Carl Auer von Welsbach	Found in the ores bastnasite and in monazite, which contain all of the natural rare earth elements; alloy used in the auto and aircraft industries.
Austrian mineralogist Carl Auer von Welsbach	Highly magnetic and used in many commercial applications; can be used to detect counterfeit paper money by showing whether the printing ink is magnetic.

Element	Symbol	Atomic No.	Year isolated
Promethium	Pm	61	1947
Samarium	Sm	62	1879
Europium	Eu	63	1901
Gadolinium	Gd	64	1886
Terbium	Tb	65	1843
Dysprosium	Dy	66	1886
Holmium	Ho	67	1879

Discoverer	Comment
American scientists Jacob A. Marinsky, Lawrence E. Glendenin, and Charles D. Coryell	Produced synthetically in nuclear reactors and accelerators; does not exist on Earth, but likely present in several stars; named for the Greek god Prometheus, who stole fire.
French chemist Paul-Émile Lecoq de Boisbaudran	Highly magnetic rare earth element; as oxide it absorbs infrared radiation.
French chemist Eugène-Anatole Demarcay	Among the rarest of the rare earths; as an oxide used to enhance the red in color computer monitors and televisions; also improves efficiency of fluorescent lights.
French chemists Jean de Marignac and Paul-Émile Lecoq de Boisbaudran	Present in compounds used to produce phosphors for color televisions and computer screens; used to detect metal weaknesses in ships and airplanes.
Swedish chemist Carl Gustaf Mosander	As a metal it is malleable and similar to lead, but much heavier; in compounds and alloys used in televisions, compact discs, and x-ray screens.
French chemist Paul-Émile Lecoq de Boisbaudran	Stable and soft metal; some isotopes absorb neutrons and may be useful in nuclear reactors; used in color televisions and compact discs.
Swedish chemist Per Teodor Cleve	Named after the Latin word for the city of Stockholm; used to color glass, but has few other applications.

Element	Symbol	Atomic No.	Year isolated
Erbium	Er	68	1843
Thulium	Tm	69	1879
Ytterbium	Yb	70	1878
Lutetium	Lu	71	1907
Hafnium	Hf	72	1923
Tantalum	Ta	73	1802
Tungsten	W	74	1783

Discoverer	Comment
Swedish chemist Carl Gustaf Mosander	Found in the minerals xenotime and euxerite of which it is an impurity; soft and malleable metal; has few uses other than in inexpensive glass and jewelry.
Swedish chemist Per Teodor Cleve	Derives from the mineral monazite; very scarce and expensive; has few commercial applications.
French chemist Jean de Marignac	Shiny and soft as a metal; used to strengthen stainless steel alloys.
Austrian mineralogist Carl Auer von Welsbach and French scientist Georges Urbain	Expensive and rare with few commercial applications; name derives from the ancient Roman name for Paris.
Dutch physicist Dirk Coster and Hungarian physicist Georg Karl von Hevesy	Shiny metal resistant to corrosion and chemically similar to zirconium; used chiefly to absorb thermal neutrons in nuclear reactors.
Swedish chemist Anders Gustav Ekeberg	Hard anticorrosive metal sometimes substituted for platinum; used in electrolytic capacitors to power cell phones and computers.
Spanish brothers Juan José and Fausto de Elhuyar	Metal with the highest melting and boiling points; sometimes called wolfram, hence its symbol; used in light bulb filaments; helps steel blades hold their sharpness.

Element	Symbol	Atomic No.	Year isolated
Rhenium	Re	75	1925
Osmium	Os	76	1803
Iridium	Ir	77	1803
Platinum	Pt	78	1741
Gold	Au	79	
Mercury	Hg	80	
Thallium	Tl	81	1861

Discoverer	Comment
German chemists Ida Tacke, Walter Nodack, and Otto Carl Berg	Extremely rare; useful in making metal alloys wear resistant, especially those used to make electrical contacts.
British chemist Smithson Tennant	Brittle metal found chiefly in ores also containing nickel and platinum; its alloys used in manufacturing electrical contacts and fountain pen nibs.
British chemist Smithson Tennant	Among the hardest anticorrosive metals; often found in ores of platinum and nickel; added to platinum to enhance hardness.
British scientist Charles Wood	Durable and malleable precious metal; mined in South Africa; commercial uses include the automobile, petroleum refining, and electronics industries; also used in cancer treatment.
Known to the ancients	Soft anticorrosive precious metal whose name comes from the Latin meaning "shining dawn"; long used as currency and in jewelry; also in electronics and dentistry.
Known to the ancients	The only metal liquid at room temperature; chief ore is cinnabar found mostly in Italy and Spain; dissolves other metals to form amalgams useful to industry; highly toxic.
British scientist Sir William Crookes	Scarce, malleable, highly toxic leadlike metal; its isotope thallium-201 is used in diagnostic medicine.

Element	Symbol	Atomic No.	Year isolated
Lead	Pb	82	
Bismuth	Bi	83	1753
Polonium	Po	84	1898
Astatine	At	85	1940
Radon	Rn	86	1900
Francium	Fr	87	1939

Discoverer	Comment
Known to the ancients	Soft, malleable metal; used since Roman times in plumbing; used to make batteries and in television and computer screens to reduce radiation; many useful alloys and isotopes.
French nobleman Claude Geoffroy, the Younger	White, brittle metal often found with copper, tin, and lead; its alloy, called Wood's metal, has a low boiling point making it useful in triggering fire alarms.
French and Polish chemists Pierre and Marie Curie	Named for Marie Curie's native Poland; formed by the radioactive decay of uranium and thorium; has the most identified isotopes of any element.
American chemists Dale R. Corsun, K. R. Mckenzie, and Emilio Segrè	Member of the halogen group and similar to iodine; first produced by bombarding bismuth with alpha particles, but results naturally as uranium and thorium isotopes decay.
German physicist Friedrich Ernst Dorn	Heavy, radioactive noble gas widely present; results from uranium decay; even present in soil; highly toxic but valuable as a cancer treatment.
French chemist Marguerite Perey	Heavy, unstable, alkali metal; produced by radioactive decay of uranium and thorium; probably less than one ounce is present in the Earth's crust.

Element	Symbol	Atomic No.	Year isolated
Radium	Ra	88	1898
Actinium	Ac	89	1899
Thorium	Th	90	1828
Protactinium	Pa	91	1913
Uranium	U	92	1841
Neptunium	Np	93	1940
Plutonium	Ps	94	1941

Discoverer	Comment
French and Polish chemists Pierre and Marie Curie	Luminescent metal and last of the alkaline earth elements; produced by decaying uranium; used in cancer treatment.
French scientist Andre Debierné	Produced by decaying uranium and thorium; 26 known isotopes; lends its name to the 14 elements that follow it, which are called the Actinides.
Swedish chemist Jöns Jakob Berzelius	Radioactive metal named for Thor, the ancient Scandinavian god of war; promising as a future source of nuclear energy; makes magnesium alloys heat resistant.
German physicists Kasimir Fajans and O. H. Gohring	Scarce and radioactive metal whose properties are little known.
French chemist Eugène-Melchior Peligot	Dense radioactive metal named for the planet Uranus; first used in nuclear fission in the 1930s; its isotopes fundamental to the operation of nuclear breeder reactors.
American physicists Edwin M. McMillan and Philip H. Abelson	Present in trace amounts naturally, but produced artificially mainly by bombarding uranium with neutrons; useful in the nuclear industry.
American chemist Glenn T. Seaborg	Radioactive metal formed by bombarding uranium with deutrons; used to power medical devices and spacecraft; essential to many nuclear plants.

Element	Symbol	Atomic No.	Year isolated
Americium	Am	95	1944
Curium	Cm	96	1944
Berkelium	Bk	97	1949
Californium	Cf	98	1950
Einsteinium	Es	99	1952
Fermium	Fm	100	1952
Mendelevium	Md	101	1955

Discoverer	Comment
American chemist Glenn T. Seaborg	Artificially made and radioactive; the alpha particles emitted by its isotope americium-241 enable the air to conduct electricity, making it useful in smoke detectors.
American chemists and physicists Glenn T. Seaborg, Ralph A. James, and Albert Ghiorso	Reactive metal produced by bombarding plutonium-239 with helium nuclei; named for Pierre and Marie Curie; useful for powering equipment in remote locations.
American chemists and physicists Glenn T. Seaborg, Stanley Thompson, and Albert Ghiorso	Produced by bombarding americium-241 with helium nuclei; scarce, radioactive, currently of little commercial use.
American chemists and physicists Glenn T. Seaborg, Stanley Thompson, Albert Ghiorso, and Kenneth Street	Produced by bombarding curium-242 with helium nuclei; its isotope californium-252 is excellent source of neutrons useful in research.
Group led by American physicist Albert Ghiorso	Radioactive and the seventh transuranium element; discovered in the debris of the first hydrogen bomb explosion in the Pacific; named for Albert Einstein.
Group led by American physicist Albert Ghiorso	Discovered in the debris of the first hydrogen bomb explosions in the Pacific and named for the physicist Enrico Fermi who produced the first nuclear chain reaction.
Group led by American physicist Albert Ghiorso	Produced by bombarding einsteinium-253 with helium nuclei and named for Dimitri Mendeleev, creator of the periodic table.

Element	Symbol	Atomic No.	Year isolated
Nobelium	No	102	1958
Lawrencium	Lr	103	1961
Rutherfordium	Rf	104	1969
Dubnium	Db	105	1970
Seaborgium	Sg	106	1974
Bohrium	Bh	107	1981
Hassium	Hs	108	1984

Discoverer	Comment
Group led by American Physicist Albert Ghiorso	Synthesized by bombarding curium-244 and curium-246 with carbon-12 ions; little is known of its properties; named for Alfred Nobel, Swedish inventor of dynamite.
Group led by American physicist Albert Ghiorso	Produced by bombarding three isotopes of californium with boron-10 and boron-11 ions; named for Ernest O. Lawrence, inventor of the cyclotron.
Group led by American physicist Albert Ghiorso	Named for atomic pioneer New Zealander Ernest Rutherford; little is known of its properties.
Group led by American physicist Albert Ghiorso	A team of Russian scientists may have synthesized this element as early as 1967; five known isotopes, but otherwise little is known of its properties.
Team of American physicists, Lawrence Livermore Laboratory and UC Berkeley	Produced by bombarding californium-249 with oxygen-18; chemically similar to molybdenum and tungsten; named for chemist Glenn T. Seaborg.
German physicists Peter Armbruster and Gottfried Munzenberg	Named for Danish physicist Niels Bohr; little is known of its properties.
German physicists Peter Armbruster and Gottfried Munzenberg	Name derived from the Latin name for the German state of Hesse; little is known of its properties.

Element	Symbol	Atomic No.	Year isolated
Meitnerium	Mt	109	1982
Darmstadtium	Ds	110	1994
Unununium	Uuu	111	1994
Ununbiium	Uub	112	1996
Ununtrium	Uut	113	
Ununquadium	Uuq	114	1999
Ununpentium	Uup	115	

Discoverer	Comment
German physicists Peter Armbruster and Gottfried Munzenberg	Produced by bombarding bismuth-209 with iron-58 ions; named for German physicist Lise Meitner, who first experimented with nuclear fission.
Team led by physicist Peter Armbruster	Presumably solid; white or gray metal; produced by bombarding lead with accelerated nickel atoms; decays after 0.0005 second.
German team of physicists at GSI, Darmstadt, Germany	Highly unstable and presumed to be a metal, but little else is known of its properties; produced by bombarding bismuth with accelerated nickel atoms.
German team of physicists at GSI, Darmstadt, Germany led by Peter Armbruster	Detected by the presence of a single atom after bombarding lead with accelerated zinc; thought to be a metal, but little else is known of its properties.
	Expected to be solid and radioactive, but has not yet been produced.
Russian physicist Yuri Oganessian	Plutonium-244 bombarded with the nuclei of calcium-48 produced this superheavy and relatively stable element.
	Expected to be solid, but not as yet discovered.

Element	Symbol	Atomic No.	Year isolated
Ununhexium	Uuh	116	1999
Ununseptium	Uus	117	
Ununoctium	Uuo	118	1999

SOURCES:

The History and Use of Our Earth's Chemical Elements: A Reference Guide,
Robert E. Krebs, Greenwood Press, Westport, Connecticut, 1998.

A Guide to the Elements, Albert Stwertka, Oxford University Press, New
York, 1998, and revised second edition 2002.

Nature's Building Blocks: An A-Z Guide to the Elements, John Emsley, Oxford
University Press, New York, 2001.

Discoverer	Comment
American scientists at Lawrence Berkeley National Laboratory	Discovered as the result of alpha decay of element 118; difficulty in duplicating element 118 may also cast this element's existence in doubt.
	Expected to be radioactive, this element has yet to be produced.
American scientists at Lawrence Berkeley National Laboratory	Three atoms may have been discovered when krypton ions were bombarded with lead, though the claim of this new element was retracted when attempts to duplicate failed.

Websites:

Jefferson Labs, Southeastern Universities Research Associates
http://education.jlab.org/itselemental/index.html

Los Alamos National Laboratory
http://pearl1.lanl.gov/periodic/default.htm

FURTHER READING

The reader might notice apparently glaring omissions here. For example, there are no biographies of some of the people to whom I have devoted entire chapters. This is because there are no recent biographies in English, and older out-of-print accounts of their lives are often inadequate. For example, I was able to locate three biographies of Mendeleev. One was a translation of a Russian book written during the Soviet era. As I read it, I frequently winced at the hero-worshiping style. The two others were published in the 1960s, when it was fashionable to use invented dialogue in biographies. One of the books went even further; the author invented events that never took place in Mendeleev's life.

There is a lot of material available on the Internet. However, I thought it better to forego listing a set of websites. The average life of an Internet link is about a year. It is higher for academic websites, but even these often turn out to be ephemeral. Much of the information available on the Internet is invaluable, but it doesn't have the permanence of printed materials.

Therefore, I decided to list a relatively small number of books that I thought would be of interest to the non-specialist reader. I haven't shied away from listing two scholarly works that contain a fair amount of technical material. These books are well enough written that the lay reader can skip the more difficult parts and read them for their engrossing accounts of the subjects' lives.

Blaedel, Niels. *Harmony and Unity: The Life of Niels Bohr.* Madison, Wis.: Science Tech., 1988. This isn't as good or as comprehensive a book as the Pais biography below. However it might appeal to readers who shun books containing equations.

Brock, William H. *The Chemical Tree.* New York: W. W. Norton & Co., 2000. A comprehensive and readable history of chemistry. This is the paperback edition. The cloth edition was published under the title *The Norton History of Chemistry.* Make sure that the difference in titles doesn't cause you to buy this book twice. That is what happened to me.

Cobb, Cathy, and Harold Goldwhite. *Creations of Fire.* Cambridge, Mass.: Perseus Publishing, 1995. A very readable history of chemistry.

Donovan, Arthur. *Antione Lavoisier.* Cambridge: Cambridge University Press, 1996. This book isn't as lively as the Poirier biography below, but it is a solid account of Lavoisier's life.

Emsley, John. *The 13th Element.* New York: John Wiley & Sons, 2000. A very lively "biography" of the element phosphorus.

Friend, J. Newton. *Man and the Chemical Elements.* New York: Scribner, 1953. This one isn't exactly a page turner. But it gives good accounts of the discovery of each of the naturally occurring elements.

Gell-Mann, Murray. *The Quark and the Jaguar.* New York: W. H. Freeman, 1994. Gell-Mann writes of his two main scientific interests, particle physics and the sciences of complexity.

Gleeson, Janet. *The Arcanum*. New York: Warner Books, 1998. An account of the life of Frederick Böttger and the discovery of the secret of manufacturing porcelain.

Greenway, Frank. *John Dalton and the Atom*. Ithaca, N.Y.: Cornell University Press, 1966. This book is more reliable than some of the biographies written during the 1960s.

Holmyard, E.J. *Alchemy*. Harmondsworth, U.K.: Penguin Books, 1957. Written more than 40 years ago, this is still the best history of alchemy.

Jaffe, Bernard. *Crucibles*. New York: Dover Publications, 1976. This is a Dover reprint of a book originally published in 1930. This history of chemistry contains more about the lives of the chemists than most such books.

Nechaev, I., and G. W. Jenkins. *The Chemical Elements*. Stradbroke, Norfolk, England: Tarquin Publications, 1997. This is a revised edition of a book originally published in 1944. The authors discuss the discovery of each of the chemical elements.

Pachter, Henry M. *Magic into Science*. New York: Henry Schuman, 1951. An excellent biography of Paracelsus.

Pais, Abraham. *Inward Bound*. Oxford: Oxford University Press, 1986. This book becomes rather technical at times. Nevertheless it is an excellent account of twentieth-century physicists' quest to understand the fundamental nature of matter.

Pais, Abraham. *Niels Bohr's Times*. Oxford: Oxford University Press, 1991. Some of the technical discussions in this magisterial book might be daunting to the lay reader. However, it contains the best and most comprehensive account of Bohr's life. The reader who skips the technical material and reads only the narrative is likely to be richly rewarded.

Patterson, Elizabeth C. *John Dalton and the Atomic Theory*. Garden City, N.Y.: Anchor Books, 1970. A reasonably comprehensive account of Dalton's life.

Poirier, Jean-Pierre. *Lavoisier.* Philadelphia, Pa.: University of Pennsylvania Press, 1996. This is the best biography of Lavoisier. It is a translation of a work published in French in 1993. Poirier revised and expanded it for the English edition.

Principe, Lawrence M. *The Aspiring Adept.* Princeton, N.J.: Princeton University Press, 1998. An engrossing account of Boyle's obsession with alchemy that becomes rather technical at times. However, the lay reader can simply skip these sections.

Strathern, Paul. *Mendeleyev's Dream.* New York: St. Martin's Press, 2001. This is a very readable account of the history of chemistry that contains a great deal of narrative about the chemists' lives. Don't be misled by the title. The chapter on Mendeleev begins on page 264 of this 292-page book.

Thackray, Arnold. *John Dalton.* Cambridge, Mass.: Harvard University Press, 1972. A good scholarly biography.

INDEX

A

Abelson, Philip H., 251
Academic Assistance Council, 185
Academy of Sciences of Turin, 150
Achilles, 69
Acidity/alkalinity indicator, 59
Actinium, 250-251
Adler, Ellen, 177
Agrigentium, 2
Air
    composition of, 97
    dephlogisticated, 81-82, 105, 116
    early experiments on properties of,
      54-55, 97
    "fixed," 97, 103, 114
    as fundamental element, 2, 3, 4
Alabaster, 23
Alchemist, The, 16
Alchemy
    Alexandrian, 4-6
    Arabic, 6-8

Byzantine, 32
Christianity and, 5-6, 10
derivation of name, 6
discoveries, 9, 70-71, 73
elixir of life, 6, 7, 9, 43-44
esoteric, 25
European, 6, 7, 8-10
founder, 184
frauds, 13-19, 63-66, 90-91
"Great Work" of, 9
laws against, 15-16, 66-67
modern practices, 24-25
Nestorians and, 5-6
Philosopher's Stone, ix, 6-7, 9, 10-
    14, 17, 18, 19, 21, 24, 45, 60-61,
    62, 64, 66, 70, 71
practical side of, 7, 9
projection, 60, 61, 63, 64, 66
punishment for failures, 17-19
satires, 16
secrecy, 9-10, 61, 78
society of adepts, 45, 63-66

spiritualism and, 66-67
theories and principles, 4, 7, 43, 56,
    57, 62, 69, 91-93
transmutation of metals into gold,
    6-7, 8, 9, 10-19, 24, 26, 43, 45,
    60, 90-91
websites, 24
Alcohol, Philosopher's Stone from, 11,
    14
Alexander II of Russia, 163, 171
Alexander III of Russia, 171
Alexander the Great, 4
Alexandrian alchemy, 4-6
Alpha particles, 176-177, 181, 182, 183,
    184, 194, 205, 208-209
al-Razi, Abu Bakr ibn Zakariyya, 7-8,
    11
Aluminum, viii, 69, 85, 86, 166, 205,
    226-227
American Philosophical Society, 107
American Revolution, 122
Americium, 252-253
Ammonia gas, 104, 149
Anaximander, 2
Anaximenes, 2, 3
Anderson, Carl, 207-208, 211
Anglican Church, dissenters and
    dissenting academies, 100, 101,
    103, 135
Annales de chimie (journal), 119-120
Antimony, 51, 60, 68, 238-239
Antiparticles, 208, 216
Apeiron, 2
Aphrodisiacs, 77
Arabic alchemy, 6-8
Arfvedson, Johan August, 225
Argon, 173, 191, 228-229
Aristotelian theory of elements, 3-5,
    35, 43, 45, 56, 57, 58, 108, 118
Aristotle, 1-2, 6
Armbruster, Peter, 255, 257
Arsenic, 7, 44, 68, 79, 232-233
Asheton, Thomas, 16

Astatine, 248-249
Astringent Mars saffron, 119
Atomic bomb, x, xi, 195, 197-201
Atomic structure. See also specific
    particles; Subatomic particles
    forces in, 203-204, 209, 210-211,
        212, 216-217
    hydrogen atom, 186-188, 205
    neutron discovery, 204-206, 208,
        209
    nucleus discovery, 182-185
    proton-electron theory, 204
    quantum theory, 185-187, 189
    radioactivity studies and, 180-181,
        182
    Thomson's "plum pudding" model,
        182
    X-ray experiments, 191
Atomic theory
    Avogadro's law and, 150-153
    Boyle's hypothesis, 57-59, 61
    Dalton's theory, 130, 138-141, 144,
        146, 147, 148, 149-150, 151, 153
    Gay-Lussac's law and, 149-153
    opposition to, 140-141, 148
Atomic weights, 146-148, 152-153, 154,
    155, 158, 165, 166, 168
Atomistic hypothesis, 58, 60
Auer von Welsbach, Carl, 241, 245
Augustus, Elector of Saxony, 19-24
Aureolus, 28
Aurora borealis, 135
Avicenna, 30, 36, 38
Avogadro, Amedeo, 150-153
Avogadro, Félice, 150
Avogadro, Philippe, 150
Avogadro's law, 150-153

B

Babbage, Charles, 142-143
Baker, Nicholas, 199

Balard, Antoine-Jerome, 235
Banks, Joseph, 95
Barium, 85, 240-241
Barium oxide, 154
Baryons, 212, 213, 220
Basil, Grand Duke of Russia, 31
Bassargin, Nicole, 159
Becher, Johann Joachim, 74, 90-91
Becquerel, Henri, 176, 180, 181
Beddoes, Thomas, 82
Beilstein, Friedrich, 169
Berg, Otto Carl, 247
Bergman, Torbern, 79
Berkelium, 252-253
Berlème, Aage, 189
Bernard of Treves, 10-14
Berthollet, Claude, 119, 149
Beryllium, 165, 205, 224-225
Berzelius, Jöns Jacob, 145-147, 148,
    227, 233, 241, 251
Beta particles, 176-177, 181, 203, 208-
    209, 217
Bewley, George, 133
Birch, Thomas, 50
Bismuth, 44, 68, 248-249
Black, Joseph, 115
Blackstone, William, 103
Bohr, Christian, 177
Bohr, Harald, 177-178, 187
Bohr, Margrethe, 186, 202
Bohr, Niels
    aid to Jewish refugees from Nazi
        Germany, 195, 196
    and atomic bomb, x, 195, 197-201
    at Cavendish Laboratory, 178-179,
        185-186, 188
    and Churchill, 200-201
    death, 202
    education, 177, 178-179
    and electron spin theory, 189-190
    escape to Sweden, 198-200
    family and early life, 177-178
    Heisenberg's visit, xi, 196-198

honors and awards, 188, 189, 201,
    202, 255
Institute of Theoretical Physics,
    188-190, 195, 201-202
and nuclear fission, x, 194-196
and periodic table, x, 191-193, 202,
    214, 220
published works, 178, 186-187
and quantum mechanics, 193-194,
    198, 209
quantum theory, 185-187, 189, 191,
    193
and Roosevelt, 200
as soccer player, 177, 178, 188
at University of Copenhagen, 186-
    187, 188-190
and World War II, x, 195, 196-201
Bohrium, 254-255
Borodin, Alexsandr, 161, 162, 171
Boron, 165, 166, 224-225
Böttger, Frederick, 17-24
Boyle, Francis, 46-47
Boyle, Richard, 46
Boyle, Robert
    as alchemist, ix, 45, 56, 60-67
    atomistic hypothesis, 57-59, 61
    chemistry interests and
        contributions, ix, 51, 52, 53-56,
        57-60, 70, 74-78, 97
    death, 67
    definition of elements, 58, 70
    early life, 46
    education and travels, 46-47
    family, 46, 50, 59, 67, 105
    and four-element theory, 56, 220
    gas experiments, 53, 54-56
    health problems, 50-51, 53, 59
    in London, 49-51, 59-60
    as natural philosopher, 52-53, 64
    at Oxford, 51-53
    and philosophical mercury, 45, 62-
        63
    phosphorus experiments, 74-78

published works, 48-49, 50, 52, 55, 56-57, 62, 63, 77, 78
religious beliefs and practices, 47-49, 51, 66-67
Boyle Lectures, 48
Boyle's law, 55-56
Brandt, Georg, 78-79, 231
Brandt, Hennig, 70-74, 76, 77, 78, 229
Brandt, Margaretha, 70
Brauner, Bohuslav, 165
Brethren of Purity, 7
Brighter Than a Thousand Stars (Jungk), 197
Bristol University, 207
Bromine, 85, 154, 165, 234-235
Bronze, 69
Brougham, Lord, 95
Brownian movement, viii, 141
Brownrigg, Dr., 80
Brunswick, Hieronymus, 33
Bucquet, Jean-Baptiste, 113
Bunsen, Robert, 86-89, 162, 235, 241
Bunsen burner, 86-87
Burnet, Gilbert, 61
Butler, Samuel (poet), 16
Byron, Lord, 103-104
Byzantine alchemy, 32

C

Cadmium, 85, 238-239
Calcination of metals, 91-92, 114
Calcium, 85, 87, 88, 192, 228-229, 257
Calcium oxide, 154
Californium, 252-253, 255
Calxes, 91-92, 104, 115
Cambridge University, 93, 96, 170, 180, 207
Cannizzaro, Stanislao, 152-153, 162
Canon Yeoman's Tale, The, 16
Canterbury College, 180
Carbon, 68, 69, 85, 165, 192, 224-225

Carbon dioxide, 53-54, 97, 103-104, 114, 140, 149
Carbon monoxide, 83, 140, 149
Carlsberg Foundation, 189
Carolian Medico-Chirurgical Institute, 145
Carra, Jean Louis, 127-128
Caustic potash (potassium hydroxide), 84, 85
Caustic soda (sodium hydroxide, 84
Cavendish, Charles, 93-94
Cavendish, Frederick, 93
Cavendish, George, 95-96
Cavendish, Henry
    death, 95-96, 100
    education, 93
    family, 93-94
    gas experiments and discoveries, 96-98, 108, 116, 117, 121, 225
    personality and lifestyle, ix, 94-95, 100, 105
    published works, 96-98
    weighing the Earth, 98-100
Cavendish Laboratory (Cambridge), 96, 178-179, 180, 184, 185-186, 188, 205
Celsus, 28
Celtium, 193
Ceramic materials, 21-24
Cerium, 145, 240-241
Cesium, 88, 89, 165, 240-241
Chadwick, James, 205, 206, 209
Chancourtis, Alexandre, 154-156, 158
Chantry, Francis, 143
Charles Albert of Italy, 151
Charles Felix of Italy, 151
Charles I of England, 46, 50, 91
Charles I of Spain, 31
Charles II of Great Britain, 74
Chaucer, Geoffrey, 16
Chemistry
    atomistic hypothesis and, 58, 60

classification of substances according to reactions, 27
combustion theory and experiments, 59, 92, 105, 108, 115, 117, 118
drug preparation, 50-51
electrical, 82-85, 103
first use of term, 27
formation of compounds, 130-131
founder, ix, 45
and four-element theory, 45
gas reactivity, 53-54, 80-82
innovations in apparatus, 146
nomenclature, 116, 119-120
and Royal Society, 45
spectroscopic analysis, 85-89
Cherwell, Lord, 200-201
Chevreul, Michel-Eugène, 152
China clay, 23
Chlorine, 69, 81-82, 147, 154, 165, 228-229
Christian, King of Denmark, 31
Christian X, King of Denmark, 188, 202
Christian Virtuoso, The (Boyle), 49
Christians and Christianity, 5-6, 10, 48
Chromium, 230-231
Churchill, Winston, 200-201
Cinnabar, 4
Clare, Peter, 143
Claude Geoffroy, the Younger, 249
Cleve, Per Teodor, 245
Cloud chambers, 208
Clozets, Georges Pierredes des, 63-66
Cobalt, 44, 78-79, 230-231
Coleridge, Samuel Taylor, 83
Color-blindness, 136
Columbia University, 195
Combustion theory and experiments, 59, 92, 105, 108, 115, 117, 118
Condensed-hydrogen hypothesis, 153
Condorcet, Jean Antoine Nicolas Caritat, 129
Constantine, 5

Cook, James, 103
Copper, 4, 59, 68, 69, 80, 87, 232-233
Copperas, 12
Corday, Charlotte, 127
Corsun, Dale R., 249
Coryell, Charles D., 243
Cosmic rays, 207-208, 211
Coster, Dirk, 193, 245
Courtois, Bernard, 239
Cousin, Jacques, 119
Crawford, Adair, 235
Crimean War, 161, 163
Cronstedt, Axel Fredrik, 79, 231
Crookes, William, 89, 165, 247
Curie, Pierre and Marie, 181, 205, 249, 251, 253
Curium, 252-253
Cyclotrons, 212

D

Dalton, Deborah, 131
Dalton, John
anecdotes and myths about, 132-133, 142-143
atomic theory, 130-131, 138-141, 144, 146, 147, 148, 149-150, 151, 153
color-blindness, 136
death and funeral, 135, 143-144, 148
education, 132-134
fame and honors, 141-143
family and early life, 131-132
as Gough's protégé, 133-134, 135
in Manchester Literary and Philosophical Society, 135, 138, 141, 143
personality and lifestyle, 135-136
as provincial scientist, 134-135
published works, 135, 138, 139, 143, 146

as schoolteacher, 132, 135, 137, 138
statue and monument, 143, 144
Dalton, Jonathan, 132, 133
Dalton, Joseph, 131
Danton, Georges, 127
Darmstadtium, 256-257
Darwin, Charles, 106
Darwin, Erasmus, 106
David, Jacques Louis, 113, 127
Davy, Humphry, 82-85, 86, 89, 141,
    225, 227, 229, 241
Davy medal, 170
de Elhaur, Juan José and Fausto, 245
de Marignac, Jean, 243, 245
Debierne, Andre, 251
Decembrists, 159
del Rio, Andres Manuel, 231
Delyanov, I. D., 171, 172
Demarcay, Eugene-Anatole, 243
Dephlogisticated marine acid air, 81-82
Desiccators, 146
Dialogue on Transmutation (Boyle), 61
Diocletian, 5
Diodati, Theodore, 50
Diogenes, 3
Dirac, P. A. M., 206-207, 213
Döbereiner, Johann, 154
Don Juan (Byron), 103-104
Donne, John, 26
Dorn, Friedrich Ernst, 249
Dostoevski, Fyodor, 159
du Mesmillet, Georges, 63-64
Dubnium, 254-255
Dumas, Jean-Baptiste, 150, 165
Dupont de Nemours, Pierre Samuel,
    124
Dysprosium, 242-243

E

Earth, as fundamental element, 3
Earth (planet), weighing, 98-100

East India Company, 64
École Polytechnique, 149
Eggs, Philosopher's Stone from, 12
Egyptians, chemical arts, 4-5
Ehrenfest, Paul, 189-190
Eightfold-way theory, 212-213
Einstein, Albert, viii, 141, 183, 187, 189,
    190, 193, 194, 206, 253
Einsteinium, 252-253
Ekeberg, Anders Gustav, 245
Electric battery, 84, 85
Electrical experiments, 82-85, 103
Electromagnetic force, 203-204, 210,
    212, 216, 217
Electrons, 176-177, 179, 204, 205, 215,
    216
    heavy, 211
    orbits, 185, 187, 203
    shell theory, 191-192
    spin, 189-190, 204, 207
Electrolysis, 84
Elements. See also specific elements;
    Table of chemical elements
    Aristotlelian theory of, 3-5, 35, 43,
        45, 56, 57, 58, 98, 108, 118
    atomic weight determinations, 146-
        148, 152-153, 154, 155, 158, 165,
        166, 168
    Boyle's definition, 58, 70
    catalog of, 223-259
    condensed-hydrogen hypothesis,
        153
    discovery of, 68-89, 223-259
    levity property, 92
    naming rights, 167
    Parcelsian theory of three
        principles, 43, 56, 57, 58, 91
    periodicity of, 153-156, 158, 162,
        164-167, 168, 170, 173, 176, 191,
        192
    philosophical, 7, 20, 45, 62-63, 66
    phlogiston theory of, 91-93, 98,
        105, 108, 114-116, 118-120

rare earths, 191, 192, 193
spectroscopic analysis, 85-89
telluric screw, 154-155
tests for detecting presence of, 59
Elements of Chemistry (Lavoisier), 82
Elixir of life, 6, 7, 9, 43-44
Elizabeth I, 46
Empedocles, 2-4
English Chemical Society, 170
Erasmus of Rotterdam, 34
Erbium, 244-245
Ernst, Duke of Bavaria, 42
Esoteric alchemy, 25
Essay on the French Disease
(Paracelsus), 39
European alchemy, 6, 7, 8-10
Europium, 242-243
Excellence of Theology (Boyle), 49
Excursion, The (Wordsworth), 134
Experiments on Air (Cavendish), 97-98

F

Faithful Brethren, 7
Fajans, Kasimir, 251
False Geber, 9
Faraday, Michael, 85
Faraday medal, 170
Faust (Goethe), 26
Faustus, Dr., 26
Ferdinand, King of Bohemia and
Hungary, 41-42
Ferdinand III, Holy Roman Emperor,
54
Fermi, Enrico, 195, 209-210, 253
Fermium, 252-253
Filter paper, 146
Finnegan's Wake (Joyce), 214
Fire, as fundamental element, 3
Fletcher, John, 132
Flowers of zinc, 119
Fluorine, 154, 165, 226-227

Foolish Wisdom and Wise Folly
(Becher), 91
Forman, Simon, 16
Foster, George Carey, 155
Four-element theory, vii-viii, 3-5, 35,
43, 45, 56, 69, 98, 108, 116-117,
118, 120, 204, 220
Four-humors theory, 35
Francium, 248-249
Frankfurter, Felix, 200
Franklin, Benjamin, 94, 102, 107
Franz Cark, Duke of Sachsen-
Lauenburg, 71
Franz Joseph, Emperor, 16-17
Frederic III, Holy Roman Emperor, 12
Frederick, Duke of Holstein, 71
Frederick I, King of Prussia, 18-19
Frederick of Wurtzburg, 15
*Free Discourse Against Customary
Swearing* (Boyle), 49
French Academy of Sciences, x, 109-
111, 113, 114, 118, 120-121, 125,
129, 141
French Revolution, 106, 125-129, 149
French Revolutionary Wars, 126-127
Fries, Lorenz, 38
Froben, Johan, 33-34, 37
Frog gold, 80
Fuchs, Klaus, 201
Fugger family, 38-39

G

Gadolin, Johan, 235
Gadolinium, 242-243
Galen, 30, 32, 34, 36
Galileo, 47, 52
Gallathea, 16
Gallium, 167-168, 232-233
Gamma radiation, 176-177, 205, 207,
208, 209
Gapon, Georgy, 173

Garibaldi, Guiseppe, 152
Gases. *See also specific gases*
　Avogadro's law, 151, 152-153
　Boyle's experiments, 53, 55-56
　Cavendish's experiments, 96-98,
　　108, 116, 117, 121, 225
　discovery, 69, 80-82
　Gay-Lussac's law, 149-150, 153
　halogens, 165
　inert, 173
　Priestley's experiments, 103-105
　reactivity, 53-54, 80-82
Gay-Lussac, Joseph Louis, 148-153, 225
Gay-Lussac's law, 148-153
Geber. *See* Jabir ibn Hayyan
Geiger, Hans, 183, 184
Geiger counter, 183
Gell-Mann, Murray, 213, 214, 220
General theory of relativity, 206
George II, 103
George IV, 141
German Cultural Institute
　　(Copenhagen), 196
Germanium, 168, 232-233, 235
Gerstorff, Hans von, 33
Ghiorso, Albert, 175, 253, 255
Gillespie, Charles C., 213
Glendin, Lawrence E, 243
Gloria Mundi, 9
Gnostics, 5
Goethe, Johann Wolfgang von, 26, 154
Gohring, O. H., 251
Goiter, 28
Gold, 4, 6-7, 8, 10-17, 24, 26, 43, 45, 61,
　　68, 70-71, 80, 166, 246-247
Goudsmit, Samuel, 189-190, 197
Gough, John, 133-134, 135
Gravitational force, 98-99, 203, 204,
　　206, 212, 216
Great Surgery Book, 41
Greenberg, Oscar W., 215
Greenup, Thomas, 135
Gregor, William, 231

Guaiac, 38-39
Grey, Lady Anne, 93
Guericke, Otto von, 54, 72
Guild of the Alfalfa, 33
Gunpowder, 122-123
Guyton de Morveau, Louis-Bernard,
　　114, 119
Gypsum, 110

H

Hafnium, 192, 193, 244-245
Hahn, Otto, 194
Halogens, 165
Hanckwitz, Ambrose Godfrey, 76, 77,
　　78
Hansen, H. M., 188
Harvard College, 48, 62
Hassium, 254-255
Hatchett, Charles, 235
Hawking, Stephen, 207
Heisenberg, Werner, xi, 190, 196-198,
　　206
Helium, 89, 173, 177, 181, 187, 191,
　　203, 224-225
Henry, Duke of Kent, 93
Henry IV, 15-16, 67
Hermes Trismegistus, 184
Hippocrates, 35, 36
Hisinger, Wilhelm von, 241
History and Present State of Electricity
　　(Priestley), 102
History of the Corruptions of
　　Christianity (Priestley), 107
Hitler, Adolph, 184, 195, 196
Hock, Vandelinus, 33
Hohenheim, Philipus Aureolus
　　Theophrastus Bombastus von.
　　See Paracelsus
Hohenheim, Ritter Georg von, 28
Hohenheim, Wilhelm von, 28-29, 32
Holmium, 242-243

INDEX

273

Home, Everard, 96
Honnauer, Georg, 15
Hooke, Robert, 54, 55
Hudibras, 16
Humboldt, Alexander von, 149
Hydrochloric acid, 81, 116
Hydrogen, 69, 81, 82-83, 84, 96-97,
    117, 118, 139-140, 146-147, 149,
    151, 153, 165, 185-187, 189, 191,
    205, 224-225
Hydrogen chloride gas, 104, 116

I

Iatrochemistry, 50
Illiad (Homer), 69
Index to the Bible (Priestley), 107
Indium, 89, 238-239
Inert gases, 173
Inflammable air, 97
International conferences, 151-152,
    162, 187
Invisible College, 51-52
Iodine, 154, 165, 238-239
Iridium, 80, 246-247
Iron, 4, 59, 68-69, 79, 80, 230-231, 235
*Iron Industry of the Urals in the Year
    1899* (Mendeleev), 173
Iron oxide, 119
Islam, 48

J

Jabir ibn Hayyan, 7, 9, 10, 11
James, Ralph A., 253
Jannsen, Pierre, 89, 225
Jars, Gabriel, 111
Johan Frederick, Duke of Saxony, 73
John George II, elector of Saxony, 71
Johnson, Samuel, 49
Joliot, Jean, 205, 206

Joliot-Curie, Irène, 205, 206
Jonson, Ben, 16
Jordan, Pascual, 190
Joyce, James, 214
Julius, Duke of Brunswick, 15
Jungk, Robert, 197
Jupiter (planet), 4

K

Kalmucks, 157
Kerosene, 85, 89
Khorassan, Emir of, 8
Kirchhoff, Gustav, 86, 87, 88, 89, 162,
    235, 241
Klaproth, Martin Heinrich, 235, 241
Klaus, Karl Karlovich, 80, 237
Kobolds, 78
Königstein fortress, 23
Kossel, Walther, 191
Kraft, Daniel, 71-73, 74-77
Kropotkin, Peter, 164
Krypton, 173, 234-235, 259
Kunckel, Johann, 71-73, 78
Kupfernickel, 79

L

Lagrange, Joseph Louis, 128
Lanthanum, 240-241
Laplace, Simon, 117, 119
Laubsoine, Marquise de, 120
Laudanum, 32, 42
Laughing gas, 83, 104
Lavoisier, Marie Anne (nee Paulze),
    112-113, 114-115, 124
Lavoisier, Antoine
    agricultural experiments and
        reforms, 123-124
    education, 109
    family and early life, 108-109

four-element theory discredited by,
108, 116-117, 120
in French Academy of Sciences, x,
109-111, 113, 118, 121
and French Revolution, x, 125-129
gunpowder production, 122-123
Marat's denunciation of, x, 121-
122, 126
marriage, 112-113, 114-115
nomenclature reforms, 119-120
personality and lifestyle, 112-113
phlogiston theory overturned by,
108, 114-116, 118-120
public service, 122-124
published works, 82, 110, 114-115,
119-120
statue, 129
as tax farmer, x, 111-113, 127-128
tax reforms, 126
theoretical work, 108, 114-120
trial and execution, x, 128-129
Law of conservation of energy, 209
Lawrence, Ernest O., 255
Lawrence Livermore Laboratory, 255,
259
Lawrencium, 254-255
Le Roy, Jean-Baptiste, 121
Lead, 4, 17, 45, 61, 68, 69, 147, 248-249
Lead oxide, 104
Leary, Timothy, 83
Lecoq de Boisbaudran, Paul-Émile,
167-168, 233, 243
Leibniz, Gottfried Wilhelm von, xi, 73-
74, 78
Lepor, Gottfried, 12
Leprosy, 38
Leptons, 216-217
Litharge, 114
Lithium, 85, 87, 154, 165, 191-192,
224-225
Locke, John, 66
Lomosov, Mikhail Vasilevich, 114

Lorentz, Hendrik, 190
Louis XVI, 106, 125
Love, as attractive force, 3
Luminescent materials, 71
Lutetium, 244-245
Luther, Martin, 40
Lutherans, 38
Lyly, John, 16

M

M theory, 219, 221
Mach, Ernst, 141
Magisteria, 27
Magnesium, 85, 166, 192, 226-227
Magnus, Albertus, 233
Malaria, 51
Manchester Literary and Philosophical
Society, 135, 138, 141, 143
Manganese, 81, 230-231, 235
Manhattan Project, 199, 200
Marat, Jean-Paul, x, 120-122, 126, 127
Marggraf, Andreas, 233
Marinsky, J. A., 243
Mars (planet), 4
Marsden, Ernest, 183, 184
Martyrdom of Theodora and Didymus
(Boyle), 49, 50
Massachusetts Institute of Technology,
213
Master Henry, 12-13, 14
Maxwell, James Clerk, 217
McGill University, 180, 181
Mckenzie, K. R., 249
McMillan, Edwin M., 251
Medicines
chemical preparations, 50-51
gases as, 82
mercury in, 27, 39, 51
phosphorus in, 77
Meitner, Lise, 194, 257

Meitnerium, 256-257
Mendeleev, Anna Ivanova (nee Popov), 170-171
Mendeleev, Dimitri
    in Bureau of Weights and Measures, 172-173
    death and burial, 174-175
    education, 159-162, 163
    European studies, 161-162
    family and early life, 158-160
    health problems, 160-161, 167, 173, 174
    honors and awards, 140, 169-170, 171, 175
    industry contributions of, 164, 168-169, 172-173
    and map of Russia, 173
    marriage and children, x, 163-164, 168, 170-171, 172
    and periodic law and table, x, 158, 162, 164-167, 168, 170, 173, 176, 191, 192, 213, 220
    personal characteristics, 157-158
    political and social views, x, 169, 171-172, 174
    published works, 163, 164-165, 166-167, 169, 173
    in Siberia, 157, 158-160
    in St. Petersburg, 160-161, 163-164
    as teacher, professor, and lecturer, 161, 162, 163-164, 171-172
Mendeleev, Feozva, 168
Mendeleev, Ivan, 158
Mendeleev, Lisa, 160
Mendeleev, Maria, 158-159, 160
Mendeleev, Olga, 159, 163
Mendelevium, 175, 252-253
Menshutkin, Nicolai, 167
Mercury (element), 7, 14, 20, 27, 39, 43, 51, 55, 56, 62-63, 66, 68, 104, 115, 235, 246-247
Mercury (planet), 4

Mesons, 210-212, 213, 220
Metallic elements, 4, 7, 43. *See also* *specific elements*
    alkalis, 147, 165
    calcination (oxidation) of, 91-92
    electron behavior in, 179
    reactivity, 80-81
Meteorological Observations and Essays (Dalton), 135
Méthode de nomenclature (Lavoisier), 119
Meyer, Lothar, 153, 167
Miletus philosophers, 1-2
Mithraists, 5
Modern Charlatans (Marat), 121-122
Modern Theories of Chemistry (Meyer), 153
Moissan, Henri, 227
Molecules, 58, 148, 151
Molybdenum, 236-237
Mosander, Carl Gustaf, 241, 245
Mu mesons, 212
Munzenberg, Gottfried, 255, 257
Muons, 211-212, 216
Muslims, 6, 7
Mysteria, 27

N

National College of Alessandria, 152
Ne'eman, Yuval, 213, 220
Neils Bohr Institute of Theoretical Physics, 188-190, 195, 201-202
Nelson College, 180
Neodymium, 240-241
Neon, 192, 226-227
Neoplatonists, 5
Neptunium, 250-251
Nestorians, 5-6
Neutrinos, 206, 208-210, 216
Neutrons, 204-206, 208, 209, 214

New Age ideas, 24
New College, 135, 137, 138
*New Experiments and Observations
Made upon the Icy Noctiluca*
(Boyle), 77
*New Experiments Phisico-Mechanicall,
Touching the Spring of the Air,
and Its Effects* (Boyle), 55
New System of Chemical Philosophy, A
(Dalton), 139, 146
Newlands, John, 155, 156, 158
Newton, Isaac, 52, 66-67, 98, 185
Newton, Mary, 180
Nicholas I of Russia, 159
Nicholas II of Russia, 173-174
Nickel, 79, 230-231, 247
Nilson, Kruss, 165
Nilson, Lars, 229
Niobium, 234-235
Nitric acid (*agua fortis*), 9, 57, 60, 69,
82
Nitrogen, 69, 85, 104, 105, 147, 149,
151, 165, 166, 224-225
Nitrous oxide, 83, 104
Nobel, Alfred, 255
Nobel laureates, 148, 181, 182, 189,
194, 195, 206
Nobelium, 254-255
Nodack, Walter, 247
Novorossisk University, 161
Nuclear fission, x, 194-196

O

Ochsner, Elsa, 29
Odling, William, 154, 155, 158
Odysseus, 69
Oersted, Hans Christian, 227
Oganessian, Yuri, 257
Oil industry, 164, 168-169
Oil of vitriol, 119
Old Nick's copper, 79

Omega-minus particle, 213
Oporinus, 32
Origine of Forms and Qualities (Boyle),
57, 60
Osaka Imperial University, 210
Osmium, 80, 246-247
Ostwald, Wilhelm, 141, 148
Owens College, 144
Oxford University, 51-53, 82, 141, 170
Oxygen, viii, ix, 59, 69, 80, 81, 84, 97,
104-105, 115-116, 117, 118, 139,
147, 149, 151, 153, 154, 165, 177,
226-227, 255

P

Palladium, 80, 236-237
Pallas (asteroid), 80
Paracelsus
as alchemist, 28, 42-44
behavior and lifestyle, xi, 32, 36-37,
42
chemical experimentation, 27, 44,
53
contributions, 27-28, 44
controversies and conflicts, 36-39,
42
death, 42
doctrine of signatures, 70
education, 29-30, 31
fame and fortune, 41-42
family and early life, 28-29, 32, 42
fantastic theories, 35-36
lampooning of, 36-37
legends and stories about, 26, 29
as physician, surgeon, and healer,
27-28, 30, 31, 32-36, 37, 38, 39-
40, 41-42
publications, 37, 39, 41
religious views and activities, 38,
39-41, 44
as teacher/lecturer, 34-36

three-principles theory, 43, 56, 57, 58, 91, 220
travels, 29-32, 39-41
Paracelsus College, 24
Pardshaw Hall School, 132
Particle accelerators, 208, 212, 218
Particle physics. *See* Atomic structure; Subatomic particles; *specific particles*
Pauli, Wolfgang, 190, 209-210
Paulze, Jacques, 112
Pelican vessel, 117
Peligot, Eugene-Melchoir, 251
Pelletan, Monsieur, 142
Penelope, 69
Perey, Marguerite, 249
Periodic law, 153-156, 158, 162, 164-167, 168, 170, 173, 176, 191, 192; *see also* Table of elements
Perkin, William, 157
Perrier, Carlo, 237
Persia, 6, 8
Petroleum Industry in Pennsylvania and the Caucasus (Mendeleev), 169
Petty, William, Lord Shelburne, 105-106, 115
Pfeffer-Ragatz spa (St. Moritz), 41
Philalethes, Eiraneus, 62
Philosopher's Stone, ix, 6-7, 9, 10-14, 17, 18, 19, 21, 24, 45, 60-61, 62, 64, 66, 70, 71
Philosophic wool, 119
Philosophical elements, 7, 20, 45, 62-63, 66, 220
Philosophical Magazine (journal), 186-187
Philosophical Transactions of the Royal Society (journal), 52, 62
Phlogiston, 91-93, 96-98, 105, 108, 114-116, 118-120
Phosphorus, xi, 59, 70-77, 85, 114, 166, 228-229
Photons, 210

Physica Subterranea (Becher), 91
Physico-Technical Institute, Berlin, 179
Pi mesons, 212
Pions, 212
Pirogov, Nicolai, 161
Plague, 40
Planck, Max, 186
Platinum, 79-80, 166, 237, 246-247
Plato, 6
Playfair, Lyon, 143
Pletnov, Ivan, 160
Plutonium, 250-251, 257
Pneumatic Institute, 82
Polonium, 248-249
Pope John XXII, 11
Porcelain, 21-24
Positrons, 206, 207, 208
Potassium, 84-85, 87, 89, 147, 154, 165, 192, 228-229
Potassium nitrate, 122
Powell, Cecil, 211, 212
Prague University, 165
Praseodymium, 240-241
Priestley, Jonas, 100
Priestley, Joseph
  in America, 106-107
  contributions and discoveries, ix, 98, 103-105, 108, 116, 227
  education, 101
  experimental research, 97, 98, 102, 103-105, 114, 115
  family and early life, 100-101
  health problems, 101
  as minister and dissenter, 100, 101-103, 107
  patrons, 105-107
  politics and persecution of, ix, 103, 106-107
  published works, 101, 102, 107
Princeton University, 170, 219
Principles of Chemistry (Mendeleev), 164-165
Proactinium, 250-251

Producibleness of Chymical Principles (Boyle), 61
Projection, 60, 61, 63, 64, 66
Promethium, 242-243
Protestant and Papist (Boyle), 49
Protons, 203, 204, 208, 209, 214, 215
Prout, William, 153

Q

Quakers, 132
Quantum chromodynamics, 215-216
Quantum mechanics, 190, 193-194, 198, 206-207, 210-211
Quantum theory, 185-187, 189, 191, 193, 206
Quarks, 214-216, 217
Quintessences, 27

R

Radioactivity, 176-177, 180-181, 182, 206
Radium, 250-251
Radon, 248-249
Raleigh, Walter, 46
Ramsay, William, 157, 229, 235, 241
Ranelagh, Lady Katherine, 50, 59, 67, 77
Rare earths, 191, 192, 193, 202
Rayleigh, Lord, 229
Recreational drugs, 83
Regnault, Henri, 161-162
Reich, Ferdinand, 89, 239
Researches, Chemical and Philosophical, Chiefly Concerning Nitrous Oxide, or Dephlogisticated and Nitrous Air, and Its Respiration (Davy), 83
Respiration, studies of, 59-60
Rey, Jean, 92

Rhazes. *See* al-Razi
Rhenium, 246-247
Rhodium, 236-237
Richter, Hieronymous, 89
Robespierre, Maximilien, 127, 149
Robinson, Elihu, 133, 135, 137
Rockefeller Foundation, 195
Roget, Peter, 83
Roman Catholic Church, 10, 40, 48, 100
Roosevelt, Franklin D., 200
Rosicrucians, 24
Royal College at Vercelli, 150
Royal Danish Society of Sciences and Letters, 201
Royal Institution, 83-84
Royal Manchester Institution, 143
Royal Society, 45, 52, 55, 75, 77, 80, 94, 95, 102, 141, 170
Rubber tubing, 146
Rubidium, 88-89, 165, 234-235
Rupecissa, Johannes de, 11
Rupert, Prince, 91
Russia. *See also* Soviet Union
    Academy of Art, 171
    Academy of Sciences, 169, 171
    Assembly of Russian Workingmen, 173
    Bureau of Weights and Measures, 172-173
    Chemical Society, 167
    Decembrists, 159
    Duma, 174
    education system, 159-160, 171
    Imperial Free Economics Society, 164
    Imperial Technological Institute, 169
    oil industry, 164, 168-169
    railway system, 160, 163, 174
    Revolution, 173-175
    secret police, 163
    women's status, 171

Ruthenium, 80, 236-237
Rutherford, Daniel, 225
Rutherford, Ernest
  and Academic Assistance Council, 184-185
  Bohr and, 179, 186-187, 188
  death, 185
  education, 180
  family and early life, 179-180
  honors and awards, 181, 182, 184
  marriage and children, 180, 184
  published papers, 181
  research contributions, 180, 181, 182-185, 204-205, 209
Rutherford, Eileen, 180, 184
Rutherford, James, 180
Rutherford, Martha, 180
Rutherfordium, 254-255

S

S matrix theory, 198
Sacrobosco, Johannes de, 11
Sage, G. B., 121
Sal ammoniac (ammonium chloride), 69
Salam, Abdus, 217
Salt principle, 43, 56, 57
Saltpeter, 122
Salversan, 27
Salzburg, 33
Samarium, 242-243
Saturn, 4
Scandium, 168, 228-229
Sceptical Chymist, The (Boyle), 56-57, 58, 220
Scheele, Carl Wilhelm, ix, 81, 86, 229, 231, 237
Schrödinger, Erwin, 190, 206
Schuler, Georg Hermann, 73
Sea salt, Philosopher's Stone from, 12
Seaborg, Glenn T., 251, 253, 255

Seaborgium, 254-255
Secret of Secrets (al-Razi), 8, 11
Segrè, Emilio, 237, 249
Selenium, 85, 86, 147, 154, 232-233
Seraphic Love (Boyle), 48-49, 50
Silicon, viii, 69, 85, 86, 166, 192, 226-227
Silicon fluoride, 104
Silicosis ("miner's disease"), 28
Silver, 4, 68, 69, 86, 90-91, 147, 236-237
Soda water, 103-104
Soddy, Frederick, 181
Sodium, 85, 87, 89, 147, 154, 165, 192, 226-227
Some Considerations Touching the Style of the Holy Scriptures (Boyle), 49
Some Physico-Theological Considerations About the Possibility of the Resurrection (Boyle), 49
Sommerfeld, Arnold, 187
Sorbonne, 31
Soviet Union. See also Russia
  atomic bomb, 199-200, 201
Special theory of relativity, 141, 183, 206-207
Specifics, 27
Spectroscopic analysis, 85-89
Spengler, Lazarus, 39
Spiritualism, 66-67
St. Petersburg Pedagogical Institute, 160, 161
Stahl, Georg Ernst, 91-92
Stanford Linear Accelerator Laboratory, 215
Starkey, George, 62
Stas, Jean Servais, 165
Stern, Otto, 187, 190
Strasbourg, 33
Street, Kenneth, 253
Strife, as physical force, 3
Strohmeyer, Frederich, 239
Strong acids, 9

Strong nuclear force, 212, 215, 216, 217
Strontium, 85, 87, 88, 234-235
Strontium oxide, 154
Subatomic particles. *See also specific*
    particles; Atomic structure
    antiparticles, 208, 216
    color charge theory, 215-216
    cosmic-ray studies, 207-208, 211
    discovery, 176
    eightfold-way theory, 212-213
    matter particles, 218
    order in table of, 213, 220
    physical laws at subatomic level,
        209
    quark theory, 214-216
Sulfur, 7, 43, 56, 68, 69, 85, 154, 228-
    229
Sulfur dioxide, 104, 116
Sulfuric acid, 9, 116, 119
Summa Perfectionis (Jabir), 7
Supergravity theory, 219
Superstring theory, vii, 217-221
Swedish Academy of Sciences, 81, 146
Syphilis, 27, 38
Syria, 5-6
Syrup of violets, 59

T

Table of chemical elements
    atomic weights and, 155, 165
    catalog, 223-259
    early attempts at ordering, 153-156
    law of octaves, 155
    periodicity, 158, 162, 164-167, 168,
        176, 202
    prediction of undiscovered
        elements, 158, 166, 167-168
    telluric screw, 154-155
Tacke, Ida, 247
Tantalum, 244-245

Tatars, 31-32
Tauons, 216
Tausend, Franz, 17
Technetium, 236-237
Telluric screw, 154-155
Tellurium, 154, 166, 238-239
Tennant, Smithson, 80, 247
Terbium, 242-243
Terray, Joseph Marie, 112, 113
Teutonic Knights, 31
Thales of Miletus, 1-2, 3
Thalhauser, Wolfgang, 41
Thallium, 89, 246-247
Thenard, Louis-Jacques, 225
Theophrastus, 29
Thirty Years War, 22, 70, 90
Thomson, J. J., 176, 178-179, 180, 182,
    184, 185
Thomson, Stanley, 253
Thorium, 147, 250-251
Thought experiments, 194
Three Papers Concerning Experiments
    on Factitious Air (Cavendish),
    96-97
Thulium, 244-245
Timaeus, 3
Tin, 4, 68, 69, 193, 238-239
Tinctures, 27
Titanium, 230-231
Torsion balance, 99
Trafford, Sir Edmund, 16
Traité élémentaire de chimie (Lavoisier),
    120
Transmutation of metals into gold, 6-7,
    8, 9, 10-19, 24, 26, 43, 45, 60, 90-
    91
Tschirnhaus, Ehrenfried Walter von,
    21, 22
Tube Alloys project, 199
Tuberculosis, 82, 101, 158, 160-161
Turgenev, Ivan, 162

U

Uhlenbeck, George, 189-190, 195
Ulloa, Don Antonio de, 79
Unification of theories, 217-221
Unitarians, 102-103, 133
University of
  Basel, 34-35, 37, 38
  California at Berkeley, 255
  Copenhagen, 177, 186-187, 188-190
  Ferrara, 30
  Genoa, 152
  Göttingen, 86, 170, 190
  Heidelberg, 86, 87, 162
  Manchester, 181
  Maryland, 215
  Moscow, 169
  Pennsylvania, 106
  Pisa, 152
  St. Andrew, 120
  St. Petersburg, 161
  Southern California, 219
  Tübingen, 28-29
  Turin, 151
  Uppsala, 145
  Vienna, 30
  Wittenburg, 71
Ununbiium, 256-257
Ununhexium, 258-259
Ununnilium, 256-257
Ununoctium, 258-259
Ununpentium, 256-257
Ununquadium, 256-257
Ununseptium, 258-259
Ununtrium, 256-257
Unununium, 256-257
Uranium, x, 165, 181, 194-195, 204, 250-251
Urbain, George, 193, 245
Urine, phosphorus extraction from, 70-71, 72-74

V

Vacuum experiments, 54-55, 72
van Helmont, Joan-Baptista, 53
van Marum, Martinus, 118
Vanadium, 230-231
Vandermonde, Alexandre, 119
Vauquelin, Louis-Nicolas, 225, 231
Vauvilliers, Jean François, 126
Venus, 4
Victor Emmanuel I, 151
Vinegar, 12
Vladimir, Grand Duke of Russia, 174
von Hevesy, Georg Karl, 193, 245
von Laue, Max
von Reichenstein, Franz Joseph, 239
von Weizsäcker, Carl, 196
Volta, Alessandro, 84
Voskresenski, A. A., 160

W

Wadham College, 52
Water
  as a compound, 98, 139-140, 149, 150
  as fundamental element, 1-2, 3, 4
Water baths, 146
Water gas, 82-83
Weak acids, 9
Weak force, 209, 212, 216-217
Wedgwood, Josiah, 106
Weinberg, Steven, 217
Wheeler, John Archibald, 194
Wien, Wilhelm, 183
Wilkins, John, 52
William and Mary College, 48
William IV, 142-143
Wilson, George, 100
Winkler, Clemens, 233
Wollaston, William Hyde, 80, 237

Wood, Charles, 80, 247
Wordsworth, William, 133-134
World War I, 184
World War II, 195-198
    aid to Jewish refugee, 184-185, 195,
        196
    atomic bomb, 195, 197-201

X

Xenon, 173, 240-241

Y

Yale University, 213
Ytterbium, 244-245

Yttrium, 234-235
Yukawa, Hideki, 210-211

Z

Ziegler, Marie, 15
Zinc, 68, 232-233, 239
Zinc oxide, 119
Zirconium, 192, 193, 234-235
Zorn, Frederick, 17-18
Zoroastrians, 5
Zweig, George, 214